YOUNG MATHEMATICIANS AT WORK
Constructing Number Sense, Addition, and Subtraction

Working with the Number Line, Grades PreK–3

Mathematical Models

YOUNG MATHEMATICIANS AT WORK
Constructing Number Sense, Addition, and Subtraction

Working with the Number Line, Grades PreK–3

Mathematical Models
FACILITATOR'S GUIDE

Antonia Cameron

Sherrin B. Hersch

Catherine Twomey Fosnot

HEINEMANN
Portsmouth, NH

Heinemann
A division of Reed Elsevier Inc.
361 Hanover Street
Portsmouth, NH 03801-3912
www.heinemann.com

Offices and agents throughout the world

Copyright © 2004 by Antonia Cameron, Sherrin B. Hersch, and Catherine Twomey Fosnot. All rights reserved. No part of this book may be reproduced in any form or by any electronic or mechanical means, including information storage and retrieval systems, without permission in writing from the publisher, except by a reviewer, who may quote brief passages in a review.

This material is supported in part by the National Science Foundation under Grant No. 9911841. Any opinions, findings, and conclusions or recommendations expressed in these materials are those of the authors and do not necessarily reflect the views of the National Science Foundation.

Cataloguing-in-publication data on file at the Library of Congress.
ISBN: 0-325-00673-3

Editor: Victoria Merecki
Text and cover design: Catherine Hawkes/Cat & Mouse
Manufacturing: Jamie Carter

Printed in the United States of America on acid-free paper
10   09   08   T&C   5   6

MATERIALS DEVELOPMENT

Mathematics in the City
City College of New York
Catherine Twomey Fosnot
Director and Principal Investigator

Sherrin B. Hersch, *Co-Principal Investigator*
Antonia Cameron, *Co-Principal Investigator*
Co-Authors, **Facilitators' Guides**
accompanying CD-ROMs

Herbert Seignoret, *Staff Assistant*
Juan Pablo Carvajal, *Staff Assistant*

Freudenthal Institute
Utrecht University, The Netherlands
Maarten Dolk
Co-Principal Investigator

Parul Slegers, *CD-ROM Developer*
Chris Rauws, *Software Developer*
Arian Huisman, *Software Developer*
Han Hermsen, *IT Supervisor*
Logica CMG
Anneleen Post, *Secretarial Support*
Nathalie Kuijpers, *Secretarial Support*
Peter Croeze, Koen Fransen, Frank Udo, *Interns*

PARTICIPATING TEACHERS

New York City Public Schools

Region 9
Jodi Weisbart, PS 3
Shanna Schwartz, PS 116
Kathryn Sillman, PS 234
Madeline Chang, PS 234

Region 10
Diane Jackson, PS 185
Brigida Littles, PS 185

City of New Rochelle Public Schools

Hildy Martin, Columbus Elementary
Michael Galland, Columbus Elementary

FIELD TEST STAFF

Mathematics in the City, City College of New York
Dawn Selnes, *Inservice Director*
Christine Ellrodt, *Inservice Staff*

Nina Lui, *Inservice Staff*
Carol Mosesson-Teig, *Inservice Staff*

NATIONAL FIELD TEST PARTNERS

Victoria Bill
   Learning Research and Development Center
   Institute for Learning, University of Pittsburgh
Cathy Feughlin
   The Webster Grove School District
   St. Louis, Missouri
Ginger Hanlon, Sarah Ryan, Gary Shevell
   CSD #2, New York, New York
Gretchen Johnson,
   City College of New York
   School of Education
Ellen Knudson
   Bismarck, North Dakota,
   Public Schools
Charlotte Stadler
   New Rochelle, New York
   Public Schools

Linda Coutts
   Columbia, Missouri, Public Schools
Cynthia Garland-Dore
   Aspen Elementary School
   Aspen, Colorado
Bill Jacob
   University of California, Santa Barbara
Judit Kerekes
   College of Staten Island, New York
Connie Lewis
   Tucson, Arizona,
   Public Schools
Wendy Watkins Thomson
   University of Missouri, Columbia, Missouri
   Project Construct, Missouri
Sheri Willebrand
   Santa Barbara, California
   County Education Office

EVALUATION TEAM

Joseph Glick, *Director*
Mara Heppen, *Researcher*
Stephanie Domenici Cabonargi, *Researcher*

VIDEO PRODUCTION

Cathrine Kellison, Roseville Video, *Producer*
Kiyash Monsef, *Co-Producer*
Jeffrey McLaughlin, *Editor*

VIDEO CREW

John Bianchi, Tami Evioni, Richard Henning, John Javakian, Michael Kelly, Anthony McGowan, Serafin Menduina, Mark Petracca, Ben Vandenboom, Richard Westlein

Contents

Overview ix
Introduction xi

Journey 1

- **THE CLASS AT WORK** 1
- **GROWTH AND DEVELOPMENT** 8

Journey 2

- **THE ROLE OF CONTEXT** 14
- **THE ROLE OF THE TEACHER** 22
- **DEVELOPING A COMMUNITY** 28
- **THE NUMBER LINE** 30
  - EPILOGUE 32

Appendix A: Children at Work: Analysis of Strategies 33
Appendix B: Handy Guide to the CD-ROM Clips 37
Appendix C: Dialogue Boxes 39

Overview

This CD-ROM chronicles how Hildy Martin and her second-grade class develop and construct a number line, initially as a model to represent a measurement situation and then as a model to represent their computational strategies.

The number line evolves from a measurement context (making labels for art work) where the children first measure different-length papers with multilink cubes and then figure out as a group where to mark their findings on a blank strip of paper affixed to the chalkboard. As a referent above the strip is a long length of cubes, strung in a repeating five white, five green pattern.

The teacher designed the problem so that the measurements marked on the strip would promote the use of landmark numbers—tens in particular—to serve as references for children as they place larger measurements on it, or as they move backward on the line. The paper strip blueprint becomes a number line; it is an *open* number line because only the numbers the children have used will be on it.

Later, in minilessons, Hildy Martin uses the open number line as a model to represent the various addition strategies of these young mathematicians. Eventually, the students begin to use the model as a tool to think with.

Introduction

Depending on the audience, the time frame for using the materials (a full-day workshop versus a semester's course work), and the intention behind their use, the CD-ROMs and corresponding books of the *Young Mathematicians at Work* series can be used in a variety of ways. They can be used with preservice and inservice teachers, with teacher educators, with parents, and with administrators. While we cannot predict all the possible uses of these materials and do not want to prescribe *how* facilitators will use them, we do envision *two kinds of journeys* for participants.

JOURNEY 1

In *Journey 1*, participants work with the first two folders of the CD-ROM, "The Class at Work" and "Growth and Development," responding to the questions and activities built into the digital environment. We view this as a *beginning journey*, where a non-judgmental form of *kid watching* is developed through repeated experiences of watching students at work during a mathematical investigation. In this journey, participants also explore the entire investigation. This includes the introduction of the context, watching the children at work over a period of weeks, and seeing different math congresses in which students share their thinking and where certain big ideas are discussed.

In *Journey 1*, participants also consider how students' thinking changes over the course of the investigation. Working in the folder "Growth and Development," they are given an opportunity to focus on individual children and examine clips of their work over time. They begin to think about some important questions: Did a student's strategy change over time? If so, how? What made it change?

As they ponder student growth and development over the course of the investigation, participants will also begin to reflect on what kinds of things affected student change. As they do this, they will begin to consider other critical aspects: How does a rich context support mathematical growth and development of students? What role does a teacher play in the development of mathematical ideas? How do a teacher's carefully-timed questions shift student thinking? How does a teacher create a classroom environment in which this kind of mathematics teaching and learning can occur? These questions are more deeply explored in *Journey 2*.

On the surface, *Journey 1* seems sequential in nature. It is important to remember, however, that while participants follow Hildy's investigation sequentially, individual explorations will not necessarily be linear. *How participants interpret this material*—the thoughts and questions they raise as they move through the digital environment—will determine what paths they take in this journey. As a facilitator, you will set the boundaries—where the journey will begin and end—but participants will move through this landscape at their own pace, with their own perceptions and inquiries. This is the power and beauty of the virtual classroom—it accommodates the needs and ideas of *all* learners simultaneously. Because learners can revisit this classroom again and again as a context for exploring teaching and learning, they have an opportunity to rethink their initial ideas, develop and deepen their own understanding, and think about important pedagogical tools.

As part of this process, participants will also be confronting contradictions that will naturally arise from *different interpretations* of what they are seeing. Highlight these to promote disequilibrium and engage participants in debate. As they interact with their peers, participants often have to confront their own beliefs about teaching and learning that surface in what they say, *their noticings in the virtual classroom*. When contradictions arise in *what people see*—as they invariably will—revisiting the CD-ROM to find supporting evidence pushes participants to distinguish between interpretation and observation.

Because they can revisit the same investigation repeatedly, there is opportunity for participants' perspectives to change. *Reviewing* allows for *reseeing;* they can *move beyond surface interpretations and dig for deeper meaning*. This is a powerful way to develop participants' mathematical and pedagogical repertoire. As one participant wrote in a reflective journal at the end of the first in a series of workshops:

> *What's puzzling me—and what I'm going to be thinking about for some time to come—is how we can all be watching the same video clip, and yet be seeing such radically different things! If we can't agree on what we're seeing—and this is just about the actions and words of two students—how can we possibly interpret what these actions and words mean?*

JOURNEY 2

In *Journey 2*, participants reenter the digital environment, but this time with a deeper and more focused set of questions. Now their explorations center on specific tools a teacher uses to support mathematizing. The different areas participants can explore are: (1) how to use context as a vehicle to support mathematical development ("The Role of Context"); (2) specific pedagogical tools that nurture student thinking ("The Role of the Teacher"); (3) how to create a classroom community ("Developing a Community"); and (4) how to develop mathematical modeling ("The Number Line"). Each of these topics has its own folder.

Journey 1

THE CLASS AT WORK

Introduction

⊙ *Columbus Elementary School is located in the central section of New Rochelle, New York—an urban area in Westchester County that borders the Bronx. It is a school characterized by large numbers of students living at or below the poverty level, many of whom are new immigrants. Sixty to 70 percent of the students are English language learners, while 70 percent of the students are on free lunch. The class is a second grade with twenty students. Hildy Martin is the teacher.*

Developing the Context

⊙ *In the following clip from a Grade 2 classroom, the teacher, Hildy Martin, introduces a measuring context. What do you think Hildy is trying to accomplish?*

The context of Hildy's investigation is measurement. In her introduction, Hildy tells the class that they will be creating art, which will be displayed at the end of the year in a *museum*. In a museum, artwork is labeled, and the labels must be precise—it looks sloppy if the labels are too long or too short. These labels for their art will be cut from strips of paper and placed below their pictures. Hildy's father has a friend who will cut the labels for them, but he needs the exact measurements of the various-sized papers that they will be using. Together the class and Hildy will make a blueprint to send to her father's friend that will have all these measurements to help him cut labels that will fit exactly. Hildy asks the students to work in pairs to measure a variety of paper with Multi-link cubes and to write their measurements on recording sheets on their clipboards. They are to distinguish the different sides of the paper by the nomenclature, "long" and "short." At each work station, she provides bins with loose multilink cubes in two colors.

FACILITATION TIP 1

When we use the word *students* we are referring to the second-grade children. *Participants* refers to the adult learners—e.g., preservice and inservice teachers.

Anticipating Strategies

⊙ *The children set to work with clipboards and bins with two colors of Multi-links. What strategies do you expect to see?*

2   Journey 1   Because the measurement context is broad, students set off to work in a variety of ways. Student strategies can include the ways they measure as well as the ways in which they determine the measurements. Some possible responses to the question *What strategies do you expect to see?* might be:

FACILITATION TIP 2

While the context is ostensibly about measurement, Hildy has another mathematical purpose—the development of the open number line as a *model*. Even though the CD-ROM is labeled, *Working with the Number Line, Grade 2: Mathematical Models* this *will not* be apparent to participants as they listen to Hildy's introduction or as they watch children's explorations in *Children at Work*. With the initial question on the CD-ROM, *What do you think Hildy is trying to accomplish?* participants' responses will vary. In a beginning journey, few, if any, will make the connection between the title of the investigation and Hildy's mathematical goals. Don't worry that they don't get this—their insights into her mathematical goals will deepen as they work with the materials. Later on, after they have worked with the whole sequence, including the minilessons, you can facilitate a discussion around their initial reflections and how their ideas about the mathematics embedded in this investigation changed over time. The idea that a context can be a vehicle for developing a model like the open number line requires a depth of understanding of the landscape of learning and of the role of context. This idea can be further explored when participants work in *Journey 2* with the folders "The Role of Context" and "The Number Line." Here the work is more advanced in nature because it requires participants to understand the big ideas and strategies (be familiar with the idea of a Landscape of Learning for mathematical development), to understand the role of the teacher and how a teacher's content knowledge affects decision making in planning an investigation, and to be able to implement an investigation within the parameters of a math workshop.

FACILITATION TIP 3

It might be helpful to have a whole-group discussion and chart participants' initial views about the strategies they think students will use. This chart can be referred to in a later discussion about the strategies students *did* use (were you surprised by what they did, if so, why, etc.).

As participants work, it might be helpful for you to keep track of the *kinds* of things they are writing. Since not all participants will realize that the question has to do with the mathematical strategies students are using to get their measurements, they might spend their entire time discussing affective kinds of things (e.g., children will be talking together, will have a hard time sharing, etc.). You will need to refocus them to think specifically about what kinds of *counting strategies* students use and how two colors of cubes put in bins by their teacher might influence these strategies.

- creating patterns with the cubes as they measure;
- using one side that has been built to make another side (either adding or taking cubes away);
- counting by ones;
- grouping cubes by color (e.g., in twos or fives or tens), but counting by ones;
- grouping cubes by color and using that structure to count (e.g., skip counting by fives);
- using more than one strategy or a mixture of strategies (counting by tens and ones);
- creating measurement tools (like a stick of ten) from cubes and using this to build other sticks of similar or different lengths;
- having difficulties when lengths are not whole units (since they're measuring with Multi-link cubes, what will they do if a measurement is 14 $\frac{1}{2}$?);
- using estimation, (if this side is 22, then that side, because it's longer, has to be about 34);
- measuring all four sides of the paper; not realizing opposite sides have the same measurement.

Children at Work

The children work in small groups. Ten groups of children set to work. Click on the pull-down menu Children at Work. *By selecting the names of the students, you can observe and study them at work. Describe the work of these groups. Describe what the students are doing and try to describe their thinking. Are the various strategies the same or different?*

The section *Children at Work* allows participants to view students as they measure different-sized papers for the art show. These clips are rich in detail and can be used by facilitators in a number of ways. Facilitation can include focusing participants on a student's actions and words (developing their kid watching) as well as having them think about what these mean in terms of a student's mathematical understanding.

Student strategies fall into three broad categories: counting by ones (e.g., Amirah and Jonathan); skip counting by ten (e.g., Lindsey and Lucero); and marking groups and unitizing (e.g., Alexander and Haley). A number of student pairs use more than one strategy to obtain a measurement (e.g., Lilly and Melissa mark and unitize, count by ones, and skip count by ten!). A more detailed analysis of student strategies is given in Appendix A.

Children at Work, Looking Back

Look back at the various ways the students worked. What strategies can you distinguish? How are the different strategies related? What are the differences between them? Reflect on the mathematical development you have witnessed. Any further noticings?

The strategies used by students are all counting strategies; these range from counting by ones to skip counting by tens. Students counting by ones and those using skip counting use the two colors of cubes placed out by the teacher. For the latter group, the cubes become a way for them to visually organize their counting. These color groupings are either in fives (Andrew and Shannon; Angelica and Diana) or tens (Lilly and Melissa; Lindsey and Lucero). There are two groups (Alexander and Haley/Emily and Janet), however, who use cubes of only one color and whose mental actions are not represented in their organization of the cubes, but in their physical actions on the materials.

Alexander and Haley's strategy evolves from working within the structure of ten (two fives makes a ten) and keeping track of the tens, to using a different color cube to mark each group of ten. Initially, their strategy is difficult to understand because their thinking is reflected not in their organization of the cubes, but in their *acting* on the materials. Their thinking evolves from counting all the cubes, to organizing this count to show how two fives makes a ten and keeping track of the groups of ten, to representing a group of ten with a different-color cube. The pink cube they place on each group of ten is a model of unitizing—that ten things (cubes) can also be one thing (one group of ten).

FACILITATION TIP 4

Initially, participants may not be able to distinguish subtle differences in how children count, let alone consider the big ideas underlying student strategies. They also may not look at children from a positive viewpoint (e.g., what a child *is* doing), but from a deficit model of observation, listing all the things they believe are important that the student *does not have* or *does not know how to do* (e.g., "He doesn't know how to count by ten" is a common response uttered by participants watching Haley in Clip 6).

Whatever participants' responses are, as they begin to observe and analyze the students in the clips in *Children at Work*, it is important to remember that there are a number of factors influencing *what they see*. These include their own views about teaching and learning, how much experience they have kid watching, and how deeply they understand what a child needs to know in order to answer the question, How many . . . ? All of these elements will surface in their responses to questions in the CD-ROM and during whole-group discussions.

In beginning discussions, it is important to focus on just kid watching. Before participants think about what big ideas underlie a particular strategy a child is using, they first need to be able to describe that strategy without processing it through an interpretation filter (see Dialogue Box A in Appendix C, page 39 for an example of how a facilitator works to develop participants' kid watching). As participants learn how to accurately describe (or imitate) students' actions and words without *interpreting or judging their behavior*, the differences between student strategies will become easier for them to distinguish.

FACILITATION TIP 5

As participants work through this section, they may need to distinguish students by name. To help with this, student pictures are located under INFO on the menu bar. Pull the menu down; student names with their pictures are located under Student Info.

Reflections on Children at Work

You have been watching second graders at work. Do you recognize such a classroom setting? Is this like the classrooms you experienced when you were a child?

In the folder "The Class at Work" participants are invited into Hildy Martin's classroom. For most people, her learning environment has no relationship to their own schooling experiences. Classrooms set up in this way (e.g., there is a meeting area with a rug, work tables where students work in groups, a wide variety of materials

4   Journey 1   that are accessible to students) will not be new for some teachers, parents, and administrators, but for others it will be startling.

FACILITATION TIP 6

It is important to note in the pull-down menu, *Children at Work*, the students are listed in alphabetical order. This list *does not* indicate a hierarchy of strategies—the strategies are not sequenced from least to most sophisticated. Indeed, Alexander and Haley, the first pair, have a strategy that is mathematically complex and extremely difficult to understand. The way Haley counts indicates the structure of his thought process, but because very little else is said by the two students as they work, participants have to rely primarily on their powers of observation; in a beginning journey this is exactly what they're *developing*.

Depending upon your time constraints and the goals for your work with participants, you may want to suggest which groups of students they view. You can focus the conversation on these students, initially working with participants on refining their kid watching, but then moving the discussion to looking at the similarities and/or differences in student strategies (what are they doing, what's the mathematics embedded in what they're doing, how are their strategies similar or different, etc.).

In order to facilitate discussion from the CD-ROM, facilitators must have an in-depth understanding of what the students are doing: being able to describe their physical actions in relationship to what is being said; to understand how their actions are connected to their thinking (what does the student's strategy mean in terms of mathematical understanding); and to know how big ideas and strategies on the landscape of learning are connected and developed by students.

As participants initially explore the CD-ROM and analyze *Children at Work*, you will be facilitating discussions by using participants' thoughts to highlight key points, confusions, and/or contradictions. Thinking of a movie within a movie might be a helpful analogy here—you are looking at a classroom and student development of mathematical ideas on the CD-ROM *at the same time* that you are working *in real time* with participants' belief systems (this includes beliefs about pedagogy and the nature of mathematics) and mathematical content knowledge. Navigating through this 3D environment means being aware of the multilayers of meaning occurring simultaneously—you may be interpreting teachers' *thinking* about students' *thinking* at the same time you are *thinking* about your own *thinking* in relationship to what teachers are saying. This *is* as complex as it sounds. While this guide can give you some suggestions on how to work with the materials, none of these can be substitutions for your own in-depth explorations and analyses of the CD-ROM and the corresponding chapters in the companion book, *Young Mathematicians at Work: Constructing Number Sense, Addition, and Subtraction*.

FACILITATION TIP 7

It is important for participants to think about counting in relationship to the development of specific mathematical big ideas. One way to bring these ideas up for discussion is to have participants focus on the similarities and differences among student strategies. Some of these are subtle and may be missed by participants who are developing their kid-watching skills, so juxtaposing specific clips may highlight and differentiate student strategies.

In this section, participants may also reflect on:

- **the role of the teacher** (some may actually wonder, When is she going to teach?)
    - the expectations for learning that Hildy Martin clearly sets (e.g., all students are investigating this problem and are expected to contribute solutions);
    - the respect she shows for students' ability to think (e.g., students are free to determine their own strategies—no hints are given for how to begin; no strategies are modeled);
- **the role of the students, who**
    - are confident in their ability to do mathematics;
    - are actively involved in solving the problem;
    - work cooperatively with a partner;
    - show respect for each other's ideas;
    - use communication to question, disagree, and defend their thinking;
    - exhibit autonomy from their teacher; they resolve their questions and differences on their own—the teacher is not drawn into the conversation as a mediator.

There are a number of possible responses here; what participants write will reflect their experiences as a learner and/or their beliefs about teaching and learning. This page might be helpful to you as you plan your next steps in how you will use the materials.

What Would You Do Next?

Imagine that you are the teacher. You have witnessed the children at work. How would you proceed? What would you focus on? Why would you do this? What do you hope to accomplish? How would this support the growth and development of the students?

Since most participants *will not* realize that the investigation is connected to the development of the open number line as a model—although the title of the

CD-ROM is *Working with the Number Line, Grade 2: Mathematical Models*—their responses to the questions above are usually connected to the measurement context. Participant responses may include:

- structuring a discussion to highlight differences in *ways* children find the answers to their measurements with a focus on counting strategies:
  - how students count their cubes;
  - how they organize their cubes into groups to help them count; and
  - which ways of counting are more efficient;
- helping students understand measurement as a linear distance of units by
  - examining different answers they have obtained (e.g., Amirah and Jonathan, who build a frame and add an extra cube onto each measurement, obtaining different answers than other students);
  - looking at *how students measure:* where they start, etc. (e.g., are the cubes lined up with the beginning of the paper?);
  - working with the idea that measurements need to be precise (e.g., Does it matter if we get different measurements? If so, why?).

FACILITATION TIP 8

A comment commonly made by teachers working with the CD-ROM who do not understand a particular student's strategy is that they will ask a student to share whose strategy *they do not understand*. Their explanation for this is that by having this student share, they [the teacher] will be then able to understand the strategy being used. What may seem obvious to a facilitator—*you are the teacher, so during this investigation and before the congress you have ample opportunity to probe a student's strategy to better understand it*—can be an elusive idea to participants.

So why might teachers structure a whole-class conversation around the explanation of a strategy they do not understand? There may be a number of factors behind this choice. One might be that the teachers themselves have little experience interacting with their students in a way that uses questions to probe, stretch, and understand thinking *as the students are working on a problem*. In their classrooms, the first occasion to listen to student strategies *is* the share—they may think that presentation is the function of having a whole-class discussion.

Another factor may be how difficult it is for participants in a beginning journey to understand the mathematical ideas behind student strategies. Some participants seize on a student's confusion, when a child's strategy is not clear or easily understood. While they perceive it as the child's confusion, the student may not be confused at all. The confusion could be their own—the strategy is unclear to them either because they cannot look below the surface of what the child is saying and doing to understand the underlying mathematical ideas *or* because they have never seen such a strategy before and have no point of reference with which to analyze it. *Not knowing what a strategy means is scary for teachers who pride themselves on being the one who knows*. For them, one way to take control of what feels like an uncontrollable situation is to have a student explain his thinking so *everyone can understand*. But this, of course, has only limited value in supporting mathematical development. Thus such conceptions at the start of the congress need to be challenged.

Building the Blueprint

> *What does Hildy do? She brings the children back to the meeting area to share their results. They construct together a blueprint for the person who will cut the strips. You will be able to watch several smaller episodes of the discussion. While building the blueprint, Hildy asks the children for the length of several papers. Make notes, especially regarding children's developing number sense and their struggles.*

After the children's explorations, Hildy brings them together to build the blueprint on a strip of paper. During this discussion, she supports the development of the open number line model in a number of ways. She

- uses a string of cubes grouped in fives as a scaffold for those students counting by ones to support development towards a use of fives and tens;
- works with and builds a system of landmarks;
- develops number relations and number *space*;
- develops computational strategies connected to number relationships (where is 66 in relationship to 46?);
- uses the measurement context to focus on *what does a number mean* (e.g., where is 66? on the cube? at the end of the cube?);

- works with counting up and counting back as strategies;
- works with common misconceptions in subtraction (what am I "taking away?" where do I start counting back?).

FACILITATION TIP 9

There are a number of ways to use *Building the Blueprint* (see Appendix C, Dialogue Box B, page 42 for an example of how this might be done). How you structure your discussion will be contingent on *your* teaching goals and the time constraints you have. Whatever your time constraints, it is a good idea to remember that many participants will be surprised by the progression of Hildy's lessons (which is connected to her teaching goal and the role of context in developing mathematical models), that she is not focusing on *how* students measure or the fact that they get different measurements, but on building a number line model to use to explore number relations. It might also be helpful to remember that there are other aspects of this congress that might be startling to participants. They may comment on Hildy's pedagogy, noticing how

- her use of questioning pushes students to think deeply;
- she uses think time to give students enough time to think;
- she uses pair-share as a teaching tool at critical math moments;
- her teaching style is nonjudgmental; students are not corrected (e.g., *right answer/wrong answer*) so that the emphasis is on thinking and sense making;
- her conscious decision making/planning around the use of two colors, measurement, blueprint, etc.

Participants may also be struck by

- the autonomy of the students, who actually challenge Hildy in several instances;
- the amount of time students spend in discussion (a typical comment made by participants is, "My kids would never sit that long.");
- the layout of the classroom (e.g., using a meeting area for whole-class discussions);
- the growth and development exhibited by students, some of whom were counting by ones, but now in this discussion are thinking in fives and tens.

FACILITATION TIP 10

It might be helpful—*before* sending participants off to view Hildy's minilesson in *The Next Day: Mental Math*—to ask them to predict how students will solve the problems. Record participants' predictions on chart paper. After they view the minilesson compare students' actual strategies to their predictions. Have a conversation about the kinds of strategies students used and highlight the similarities and/or differences between these. (It might also be a good idea to make a mental note of the kinds of strategies participants predict [e.g., do they predict only splitting strategies and/or the standard addition algorithm?]. If their understanding of addition strategies is limited, you may want to spend some time working on addition with them to broaden their own computational strategies [e.g., doing mental math addition strings with them].)

The Next Day: Mental Math

The next day, Hildy Martin starts math workshop with a minilesson. In this minilesson students solve some addition problems. How does Hildy use the paper strip and the line with the cubes to support students' solutions?

In this minilesson, Hildy works with addition, but stays within the context of measurement and building a blueprint for her father's friend. She tells the students she has found other papers at home (one measures 19 cubes; the other looks about the same size and when she puts the 19 cubes against it, it is just 1 more). She records this measurement as the beginning of her string of problems, 19 + 1. Students immediately respond that the length of the other paper is 20.

Hildy tells them that she has also been thinking about what would happen if they combine papers. For example, if she combines two pieces of art paper, how long would the strip for the label need to be? What if she combined the paper that was 19 cubes long with another paper that was 21 cubes long? She records the new problem below what she has already recorded on the board:

$$19 + 1 = 20$$
$$19 + 21 =$$

Three students share different strategies, which are represented on the board by Hildy. Since she is working to develop the open number line as a model, she grounds the students' solutions in the context of the strip they've constructed with the other measurements (see *Building the Blueprint*). Even though the students do not necessarily use the blueprint (the paper strip with the measurements marked on it and the line of cubes organized in groups of five that is above it) as they solve the problem, Hildy brings them back to it as a way to reinforce the work they have done building number relations. Her questions (e.g., "Well, where would 21 go on this strip?") keep the conversation rooted in the context where numbers have a specific meaning connected to measurement. Her representation of their thinking on the paper strip with jumps to show their thinking

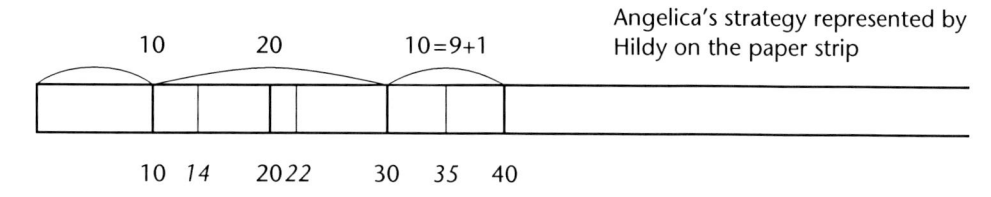

Angelica's strategy represented by Hildy on the paper strip

```
    10      20    10=9+1
  ⌒    ⌒    ⌒   ⌒
 |  |  | |  |  |   |
 10 14  2022  30  35  40
```

is a bridge to the work she will be doing with the open number line in future minilessons.

The three student strategies for 19 + 21 are shown below.

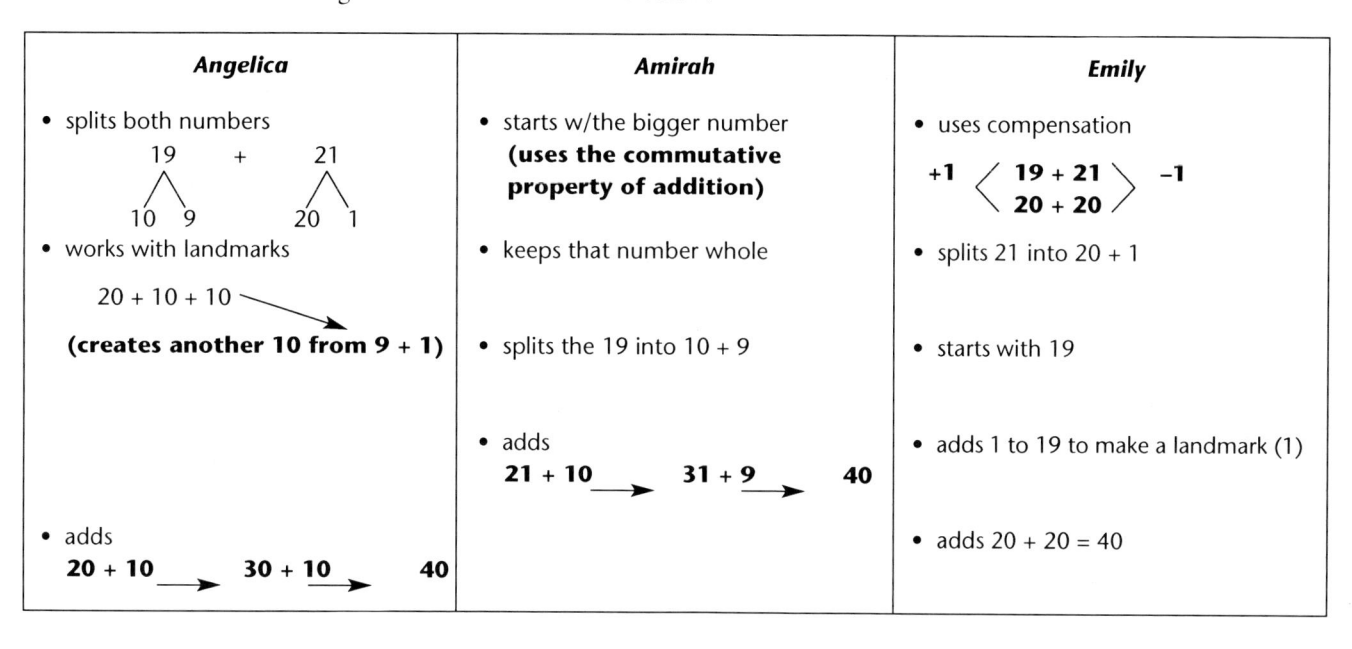

Angelica	Amirah	Emily
• splits both numbers $\quad 19 \quad + \quad 21$ $\quad /\backslash \quad\quad /\backslash$ $\,10\;\;9\quad\;\;20\;\;1$ • works with landmarks $20 + 10 + 10$ **(creates another 10 from 9 + 1)** • adds $20 + 10 \rightarrow 30 + 10 \rightarrow 40$	• starts w/the bigger number **(uses the commutative property of addition)** • keeps that number whole • splits the 19 into 10 + 9 • adds $21 + 10 \rightarrow 31 + 9 \rightarrow 40$	• uses compensation $\quad\;\;+1 \;\langle\; \begin{matrix}19 + 21\\20 + 20\end{matrix} \;\rangle\; -1$ • splits 21 into 20 + 1 • starts with 19 • adds 1 to 19 to make a landmark (1) • adds 20 + 20 = 40

Several Days Later: Mental Math

◉ *Several days later, we see the students during another minilesson. How has the representation of the students' thinking evolved?*

In this clip, Hildy is working with another addition minilesson. The problem is 43 + 20. While the cube line is still available as a tool on the board, the *blueprint* is no longer there. Now Hildy is using the *open number line* to represent student thinking. She does not refer them to the cubes (e.g., Where would 43 be?) as she represents their strategies; they are now working with the number line as a model to represent *computation strategies*.

The two strategies shown are:

Melissa	Janet
• Starts with 20 • Splits 43 into 40 + 3 • Adds 20 + 40 + 3 Hildy's representation on the open number line: $\quad\;\;\;\;40\quad\quad\quad\;\;3$ $\;\;\frown\quad\quad\quad\;\frown$ $20\quad\quad\quad\;\;60\;\;63$	• Starts with 43 • Splits 20 into 10 + 10 • Adds 43 + 10 + 10 Hildy's representation on the open number line: $\quad\;\;10\quad\quad\quad\;10$ $\;\;\frown\quad\quad\quad\frown$ $43\quad\quad\;\;53\quad\quad\;\;63$

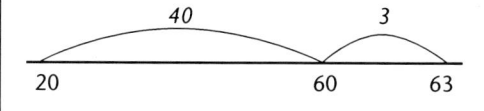

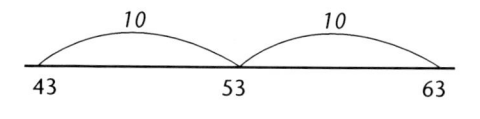

Journey 1   *Backburner: A Moment for Further Reflections*

◎ *This is the last page in the folder, "The Class at Work." However, you may have other questions on this topic that you would like to investigate. Go to the TOOLS menu above and add them to your backburner notes.*

The *Backburner* page is a tool for *both* participants and facilitators. As participants work with the materials, they may raise many questions that may not be easily or immediately answered. The *Backburner* page offers them a place to keep their questions for another time. These may be answered or evolve as they work more deeply with the CD-ROM.

The digital learning environment can be the context for further investigations, which can be rooted in the learner's own questions. A facilitator can use participants' questions to form study groups; participants with similar questions (e.g., How do different groups of students use the recording sheet provided by Hildy? Are some students more vocal than others? What are the power relationships in pairs? Is one child more dominant? etc.) can research this. (Some additional reading materials are provided under INFO in the menu bar.)

For a facilitator, participants' questions can be a window to their thinking, offering invaluable insights into beliefs. Yet in a discussion questions might very well derail the focus. A beautiful way to validate the importance of individual ponderings but *not* let these alter the flow of discourse is to encourage the use of the *Backburner* page (e.g., What a wonderful question! We won't be able to think about it right now, but if you put it on the backburner, perhaps we can think about it at another time.). In looking at questions recorded on the *Backburner* page, a facilitator can also see the evolution of a participant's thinking. It is helpful to note: Do their questions change as they work with the materials? If so, how?

GROWTH AND DEVELOPMENT

Investigating Learning

◎ *In this section you can take a close look at a few of the students over time. Describe the mathematical development—the learning—that you notice.*

In this section of the CD-ROM, the video files are organized by *child* rather than by time as in the first folder. The files are organized in this fashion to provide participants with different learning opportunities than they experienced in their prior work with the CD-ROM. Precisely because the database is organized by child, participants have the opportunity to investigate and assess an individual child over time—to try to determine what the child knows, does not know, or is in the midst of constructing.

Learning how to assess children is a critical step in participants' journeys because how they view children will inform how they teach them. Being able to carefully and clearly describe children, to recognize the importance of providing rich details as evidence of what they are saying, to know what this evidence means in terms of development, are essential tools in teaching mathematics.

In their own practice, teachers often develop this as they are teaching, but it is enormously difficult to do because of time constraints and because of the number of things they are juggling simultaneously (assessing individual children, thinking about each student in relationship to other learners in the classroom, using this information to plan and teach in ways that support the development of all learners, etc.). As the database provided in this folder is used by participants, there is a potential for them to develop and hone their powers of observation and assessment because now they have the luxury of focusing not only on one child, but on that child over time.

Because the database allows participants to view students over time, the initial description they write of a child may not be the final portrait they paint. Details can be added as they view succeeding clips, and their later observations may make them question their first impressions. Thus, another potential change can occur as participants work with the children over time: they can begin to question how they view children and in so doing, alter the way they see.

As participants raise questions about what they thought they knew about a child, they may also begin to raise questions about themselves as observers (e.g., How can I change the way I view children so that I include more details? Why did I focus on these details and not on others? Why did I think this detail was important? How can I know which details to include? Are some details more important to include than others?). As they share their portraits with other participants, they may note differences among their observations, which may also lead to another kind of self-reflection (e.g., Why did I focus on the negative aspects of a child when other people have talked about more positive kinds of things?).

They may also raise questions about what they would like to ask the child, or what activities they would engage the child in next in order to gain more clarity of the child's mathematical development. Raising these questions and then determining what they would do to find out gets right to the heart of the kind of thinking a teacher must do in the moment of teaching. Here participants have the opportunity to develop this ability.

The purpose of "Growth and Development" is twofold: (1) to develop the ability to assess children (e.g., to describe a young mathematician—in a sense to *paint a portrait* of an individual child's strategies and struggles as s/he constructs, or is on the cusp of constructing, strategies and big mathematical ideas around number); and (2) to analyze and document the mathematical *growth and development over time* of individual students. The paper tool in the pull-down menu is excellent to use here.

Activity One: Painting a Portrait. The act of painting with words a mathematical portrait of students in this class is an assignment that may help to move participants beyond the façades of the children and to enable them to know them better. Because the ways teachers view students affect the kinds of interactions they have with them, the kinds of questions they ask them, the curriculum that they plan for them, and so on, developing how they look at children is a necessary step in their journeys. For this activity, Haley, Melissa, and Shannon present challenging subjects to study:

Haley is often remembered by participants for his counting strategy (Clip 6) that evolves into a model of unitizing (where he uses pink cubes to mark the groups of ten). Because this clip is memorable (especially if you have used it in a kid-watching activity where participants, who initially analyzed Haley's strategy as a "low-level strategy," "counting by ones," or "not being able to skip count without Alexander's help,"

Growth and Development **9**

ACTIVITY 1   Painting Student Portraits

One way to work with developing participants' observational powers is to pick a child for them to paint *before* you begin working with them on "Growth and Development." Let them explore the clips of this child and write about his development. It is a good idea to select a child (e.g., Josué) who will not be used for later assignments.

One possible way to do this is to have participants suppose that Josué is a child in their classroom and that they need to write a summary of his mathematical development for his report card. The video clips of Josué will form the basis of their assessment. Leave the assignment open enough so that participants' writing will reflect how they would do this in their own classrooms. As you read their responses, think about how they describe Josué's mathematical development. Are their observations

- superficial (*Josué is measuring the yellow paper with cubes.*) or specific (*Josué uses only black cubes as he measures the yellow paper. Though we don't hear him count, his organization suggests that he will count his cubes by ones.*)?
- positive or negative (framed in terms of what he *can do* or what he *cannot do*)?
- focused on affective kinds of behavior (*He shares the cubes and works well with his partner, but is very quiet in his interactions.*)?
- a synthesis of all these details into a portrait of Josué's mathematical development?

At the conclusion of your work with "Growth and Development," return to this original assignment and ask participants to rewatch Josué's clips and critique their original portrait. Do they agree with their initial observations? How would they alter the portrait? Finally, have them reflect on what might have affected the changes in how they are now seeing Josué.

TECH TIP 1

For technical Information on how to use the Paper tool, refer to the Help file under the Tools menu.

make major shifts in their own understanding of what he is doing in this clip), participants can *fix* Haley in their mind as a *wunderkind*. When they do this, they lose sight of Haley as a learner who is still in the process of developing other mathematical ideas. Though there are only three clips in this section, these will help participants paint a more realistic portrait of Haley. He

- explains, in answer to Hildy's question, "Where should I mark 66, at the cube or at the edge of the cube?" that 66 is on the cube. When questioned by Hildy, Haley says that it is the same (e.g., on or off the cube) and therefore, it does not matter where she marks 66 on their blueprint (Clip 40); and
- recognizes, after Hildy reintroduces the context (the labels they are making for the art show that will be cut by her father's friend), that 66 should be at the edge of the cube (Clip 44).

Lucero may be a child participants overlook because she is not always able to explain her thinking. In the whole-class discussions, her contributions to the ideas being considered by the community can be overshadowed by her more vocal peers. Seeing the sequence of clips may surprise some participants who might have developed only a sketchy portrait of Lucero from their previous work with *Children at Work*. Lucero

- skip counts by ten, but misses one stack of ten hidden under Lindsey's clipboard which gives her an incorrect measurement (Clip 5);
- recognizes that they do not have to go back to zero to place a number on their blueprint and that they can place 66 by using 46 as a starting point. Though she has difficulty articulating how she would do this—she does mumble "56," which indicates that her strategy is to count by ten from 46 to 66—this is the first time in the discussion that a child has used the number line they are developing as a tool to think with (Clip 26);
  - disagrees with Haley that 66 can be marked either on or at the edge of the cube and says that it has to be marked at the edge because that is where Hildy put the line for 66 on the blueprint (Clip 41); and
  - as Hildy demonstrates Angelica's strategy to find 80 (go back 3 cubes from 84), Lucero is visibly excited, but is not given a chance to articulate her own strategy (Clip 46).

Melissa can be an example for participants of a child's use of a consistent strategy—working with ten as a landmark—in a variety of situations. Melissa

- groups her cubes in an alternating pattern of ten and uses this quantity to skip count (although she and Lilly check their count in a number of different ways, e.g., counting by ones, counting by ones to ten) in Clip 7;
- proves to the class how she knows the chart paper's measurement is 46. She puts two groups of five together on the bead string and skip counts, "10, 20, 30, 40" and adds on one group of five and one more to get 46 (Clip 39); and
- changes the problem, 43 + 20, into (20 + 20) + 3, and adds the landmarks (Clip 70).

In "Growth and Development," participants can look more deeply at students as individuals, focusing on their strategies during the investigation and comparing these to their later work in *Building the Blueprint* and in the ensuing minilessons. They can reflect on a student's initial strategies, and compare those to his thinking over time. Did the student's thinking change? If so, what supported that change? Was it

- in response to another student's thinking?
- a well-timed question from Hildy?
- the use of context and its development over time?
- the use of the materials during the *congress* (e.g., the line of cubes in alternating groups of five with the paper strip below it)?
  - Hildy's choice of numbers as they build the blueprint?

SAMPLE PORTRAIT: LUCERO

When we were given this assignment and asked to select from three children, I picked Lucero because I couldn't remember anything significant about her. Before watching these clips, I went back to my notes from *Children at Work* because I thought that even if I didn't remember her vividly, I would have written about her and maybe this would jog my memory.

As an aside here—to give you an idea of how shocked I was by what I had missed—I pride myself on being a good kid watcher. While I don't always take notes in my own classroom, I make mental images of children at work and in their interactions with other children, and use these as the basis of my assessments. Imagine my surprise when I couldn't remember Lucero in detail and found that my notes were so sketchy that I still could not paint *her portrait*!!

When I went back and watched the video clips in this section, I was stunned by what I had missed. How did I not originally see how much she contributed to the mathematical life of the classroom community? How did I miss that it was Lucero who suggested that they could place 66 on the blueprint without going back to the beginning of the cube string to count? This was a big moment in that class. In my notes, I had recorded this under Emily!! Why did I think it was Emily who had initiated that conversation? Was it because Emily was much more vocal and could more clearly articulate her ideas? (I was so struck by the fact that I had missed this detail in my original observations that I had to go back and rewatch *Building the Blueprint* to check.)

So Lucero began what I think of as a shift in the mathematical discussion in Hildy's class because they were moving away from using the cube string as a support (Clip 26). Actually, I think because they had spent so much time using the cubes, moving away from them was especially difficult at this moment because the ten Lucero was jumping by was no longer visible (i.e., all along, ten was connected to five green and five white cubes; now the ten was split into 4 + 5 + 1). So ten now becomes an abstraction—a tool to think with.

Lucero also recognized that there was something strange in Hildy's question, "Where should 66 go, on or off the cube?" since Hildy all along had been placing the mark at the end of the cube (Clip 41). She was that observant. Hildy had been marking the measurements all along and using the edge of the cube to mark these. The mark for 66 was up on the blueprint already, so there's something ludicrous (not mathematically, but realistically) in Hildy's question. Perhaps, if Hildy had realized this, she could have asked Lucero, "what if I made a mistake? Is there some way you could prove to me 66 should be at the edge of the cube?" to really probe the depth of her understanding.

Then in Clip 46, I was struck by her enthusiasm as Hildy was working with the strategy of where to place 80 in relationship to 84. Lucero was so involved in what was happening mathematically, she was almost bursting at the seams to share her thinking. I was disappointed actually that she did not get to do that because I interpreted her shaking of her head as Hildy counted back three cubes as an indication that she knew it had to be a different answer.

So here is Lucero starting to come to life before my eyes. Mathematically Lucero has a lot to contribute to the class and she is eager to do so. What holds her back sometimes is the fact that she is not always able to explain her thinking clearly.

Creating this portrait has made me reflect on my own teaching and powers of observation. I have some questions now for myself when I reenter the classroom. Which children do I notice? Am I able to describe all my students or do I take time to notice certain students, who, for whatever reason, jump out at me. I have begun reflecting on myself as an observer of children and know that this experience has helped me become more aware of the journey I need to take in order to be a better painter.

"Growth and Development" is a pivotal section in the CD-ROM learning environment. In pondering the questions Why do students' strategies change? and What makes them change? participants are confronted with pedagogy connected to a constructivist view of teaching and learning and to the idea that doing mathematics is a journey in which *mathematizing* occurs in each learner at increasingly complex levels and in different ways. In this CD-ROM, Hildy supports this learning through a very carefully thought-out sequence of activities, whose overarching goal is the development of the open number line model.

Activity Two. Ask participants to investigate a child over time and write a paper that documents and describes how the child's thinking changed during that period. Because of the range between their entry points into the measurement investigation and where their journey ends, Amirah, Angelica, Emily, and Janet are interesting subjects to study, although you might want to suggest that participants focus on just one of them. Many things have contributed to their growth and development: the context and the materials, interactions with others, constraints and shifts in the investigation, questions from the teacher, and so on. The idea here is that it is *the possibilities for learning* within all these factors and not the factors in and of themselves. (Sample dialogue is provided in Appendix C, page 45.

Not all the children in Hildy's class make huge developmental shifts in their thinking (nor would it be realistic to expect they would!). There are, however, a number of children whose strategies shift over the course of the investigation and minilessons. Listed below is an analysis of the growth and development of four of these students.

FACILITATION TIP 11

As they explore "Growth and Development," participants will recognize shifts in students' thinking. That student strategies can change so quickly (over the course of the investigation and minilessons) will be surprising to some—especially if they think teaching and learning are about steps provided by the teacher and practiced by the student until mastered; the recognition of this belief is heard in a commonly uttered phrase, "They get it."

As you are developing and working with the idea of a landscape of learning, this section of the CD-ROM will help participants understand the landscape as a metaphor for thinking; there are many paths and journeys, but which ones students take is determined by them. Thus individual journeys cannot be predicted. (Would anyone watching Amirah count by ones as she built a frame in *Children at Work* have predicted her facility in working with ten as a landmark both in *Building the Blueprint* and later in the minilesson?) Students, as they explore and reflect, sharing their ideas and listening to those of others, can make huge leaps in their thinking.

This is a big idea for teachers to grapple with—that student thinking is not necessarily linear, that learning can come in spurts, and that learners can shift their thinking in profound and unpredictable ways. If this is true, important questions for teachers become

- what kinds of experiences support student growth and development?
- what series of activities will be rich enough to support and maximize mathematizing?
- how should this sequence of activities be structured?
- what is the mathematics embedded in this investigation?
- what will student strategies be?
- what is my role in facilitating student understanding?
- what materials should be provided?
- how does communication affect student thinking?

Amirah. Over the course of the investigation and ensuing minilessons, Amirah's strategies evolve from

- counting by ones to figure out the measurement of the art paper (Clip 9) to
- using ten as a landmark to
  - figure out where 22 should go (In Clip 16, she unitizes ten "Those two [two groups of ten] would be 20") and adds on 2 more to get to 22 ("and another 2 [add to the 20]");
  - solve an addition problem, 19 + 21. (Her strategy in Clip 55 is to change the problem into 21 + 10 + 9. Here, ten is not just connected to place value and splitting, but becomes a tool to think with.)

Angelica. In *Children at Work*, Angelica is seen working with cubes that are organized in groups of five, but not using those groups to count. Her strategy evolves from

- counting by ones as she measures the art paper (Clips 72–75) to
- developing a system of tens when she
  - figures out where to place ten on the blueprint; she says, "The green and the white," i.e., five green cubes and five white cubes make ten (Clip 32);

- splits the numbers in the addition problem 19 + 21 into 20 + 10 + 9 + 1. In this strategy, Angelica is working with some big ideas in addition: the commutative property and place value. She also has developed two landmark strategies: splitting and making tens (Clip 54).

Emily. There are many clips of Emily to choose from. In *Children at Work*, Emily is seen breaking a pink cube train used to measure the purple paper into two groups of ten and two more (Clip 10). This strategy shows that ten, thinking with ten, is a tool for her to figure out the total length ("Here's 10, 20, 22"). In *Building the Blueprint*, Emily's strategy evolves from

- working with the structure of the cube string to place numbers on the blueprint—she skip counts by fives or tens, but always goes back to the beginning of the cube string to place numbers (Clips 33, 36, 48) to
- using the blueprint as a tool to think with. (This dramatic shift occurs in Clip 28 when she is trying to figure out where 66 would go in relationship to 46 and 84. Emily says, "You can start at 46, take a jump of ten to 56, and then make another jump of 10 that would get to 66.")

Other important strategies Emily is seen using are:

- using ten as a landmark tool ("take a jump of 10") in addition;
- using compensation in Clip 56 to solve the addition problem 19 + 21. (She changes the problem into 20 + 20 [(19 + 1) + (21 − 1) = 20 + 20].)

Janet. In *Children at Work*, Janet is seen working with Emily, who takes a lead role in developing the strategy they use. While Janet works in tandem with Emily, who separates 22 cubes into 2 groups of ten with 2 leftovers, it is not clear how much of this strategy she owns (Clip 10).

In *Building the Blueprint* (Clips 45 and 31), Janet is seen thinking about two problems that are closely related: *Where is 80?* (in relationship to 84) and *Where should it be marked on the number line?*; and *Where is 60?* (in relationship to 66). Her thinking here evolves from

- confusion about how many cubes she needs to go back from 84 to 80 to a
- clear explanation of how she knows 66 to 60 is 6 cubes away. Her language shows an understanding of place value, that no matter what direction you go (backward or forward), 66 is 6 cubes from 60.

In the second minilesson (Clip 71), Janet solves 43 + 20 by

- keeping 43 whole
- splitting 20 into two tens
- and taking jumps of ten (using ten as a landmark tool for adding). Here she clearly uses the number line as a tool to think with.

Backburner: A Moment for Further Reflections

This is the last page in the folder "Growth and Development." However, you may have other questions on this topic that you would like to investigate. Go to the TOOLS menu above and add them to your backburner notes.

See *Backburner* notes, page 8.

Journey 2

In *Journey 1*, participants visited Hildy Martin's classroom to experience the full range of a math workshop, which included an investigation and ensuing minilessons. In this journey, participants observed the students at work and pondered the mathematical big ideas connecting and underlying their strategies. As participants watched Hildy's whole-class discussion, they were given an opportunity to consider the ways in which she worked with students' struggles and confusions and scaffolded discussions to support the development of the number line model. They were also able to explore the ways in which individual students' mathematical ideas developed over the course of the investigation and consider what affected these changes ("Growth and Development").

In *Journey 2*, participants revisit the CD-ROM, but this time they view it through a different lens. Whereas their initial journey was broad in scope, their work now can be more highly focused. In *Journey 2*, there are four new folders to explore: (1) "The Role of Context" in which participants can begin to work more deeply with context and think about the ways in which Hildy uses it as a didactical tool to support mathematizing; (2) "The Role of the Teacher," where they can examine specific aspects of Hildy's pedagogy that contribute to students' mathematical growth and development; (3) "Developing a Community," where they can focus on ways she develops a mathematical community; and (4) "The Number Line," where they can listen to an interview in which Hildy discusses her plans for the investigation, and clip supporting evidence from the class to build a timeline of how that plan unfolds.

THE ROLE OF CONTEXT

Introduction

◎ *In this folder, about the role of context, you can investigate how Hildy uses context to support development, and how its use enables children to realize what they are doing.*

Hildy's Use of Context

◎ *Hildy uses a measurement context to support the development of the open number line. How does she accomplish this? What progression do you see from her initial investigation to the eventual use of it for mental math?*

Hildy develops the open number line model through a series of activities that are rooted in a measurement context. Participants might have peripherally thought about Hildy's use of context in *Journey 1*, but their primary focus was on analyzing children's strategies and thinking about how she used these in whole-group discussions.

Journey 2, "The Role of Context," invites participants to shift their perspective to examine how Hildy's didactical use of context supports not only the development of landmark strategies and important mathematical ideas, but also a mathematical model: the open number line. Although participants may be able to delineate *how* Hildy uses the measurement context (the structure of her investigation, when she shifts the context, etc.), being able to explain *why* she is doing this (the mathematical reasons underlying her use of context) will be a much more difficult task.

How does Hildy structure the measurement investigation? In the initial investigation, children measure different-sized papers that will be used in the art show. Hildy carefully explains the reason they need these measurements—they will be creating labels for each piece of art and, because this artwork will be displayed in a museum and must look professional, the measurements must be exact. Children set off to work with a seriousness of purpose, determining their own solution methods.

FACILITATION TIP 12

In order for you to keep participants grounded in context, it is important that *you* are clear about Hildy's didactical uses of context. In general, she uses context in two important ways: (1) to support development through the use of constraints or potentially realized suggestions; and (2) to help children realize what they are doing.

FACILITATION TIP 13

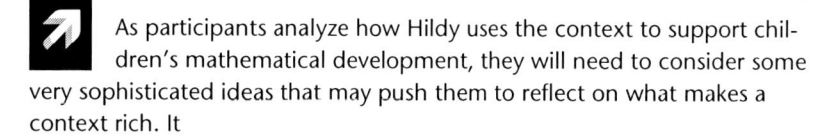

As participants analyze how Hildy uses the context to support children's mathematical development, they will need to consider some very sophisticated ideas that may push them to reflect on what makes a context rich. It

- is open-ended enough to meet the needs of all learners in a classroom, but deep enough to sustain their ongoing mathematical journeys;
- allows students sufficient time to work at their own pace;
- can be used as a teaching didactic to
  - support mathematical development; and
  - work with students' confusions so that the solution to their struggles can be realized within the situation.

In the next activity, Hildy brings the children together for a whole-class discussion and develops the blueprint for her father's friend. While she continues to work within the measurement context, she subtly moves students beyond it. This is done through a careful scaffolding of number relationships that begins with the landmark ten. This is no fluke—Hildy precut the dimensions of the paper to be measured by the students in order to ensure that certain numbers would come up for discussion. Thus she is building a system of number relations within a measurement context, supporting the development of some big ideas about number and measurement simultaneously.

In measurement, a big idea is that units are iterated and the measurement includes the entire distance measured. This idea comes up for discussion with the question *Where is 66—on the cube or off the cube?* Students have misconceptions about this question. This is resolved for Shannon when Hildy brings them back to the context—66 has to be the entire distance (to the edge of the cube and not on it), otherwise the paper strip for the art show would be too short. In this part of the investigation, Hildy jumps in and out of the context, bringing it back to the discussion as needed. It is the perfect tool to refocus student thinking and to work through their confusions and misconceptions.

In the discussion *Where is 80?* Hildy uses the blueprint and cube line to highlight confusions that are common in subtraction. Janet suggests that, in relation to 84, 80 is 3 cubes back. Hildy models Janet's thinking with the cubes and there is puzzlement on the child's face as she grapples with the fact that her instructions and the cube line do not match. Janet is left with a cube line that clearly shows 1 more than 80, which contradicts her thinking. Because her confusion can be resolved within the measurement context—the blueprint they are creating is connected to the cube line where 80 cubes can be delineated through the repeated groups of 10—her thinking shifts.

What underlies Janet's struggle are ideas that are problematic for many students in subtraction—What [how many] do I go back? Where do I start counting? Within a measurement context, where building a blueprint is connected to a cube line marking ten as a landmark, and where students are working with this knowledge, 84 clearly has 80 as its root. Although Janet initially struggles with mentally working with the relationship of 80 to 84, when asked a similar question by Hildy ("where is 60?"), she knows that the distance between the two numbers (66 and 60) is a measurement of 6 cubes. An important idea Janet works with is that the distance between those two numbers does not change whether she goes forward or backward on the cube line.

In the first minilesson, Hildy works with addition, supporting and scaffolding student thinking by continuing to use the blueprint (with the measurements previously placed on it during the math congress) and the cube line above it. The two problems (19 + 1 and 19 + 21) are grounded within the measurement context.

In the first problem, Hildy compares two papers that are almost the same size. She has measured one of the papers, and it is 19 cubes long. The other paper, when she aligns it, is only one cube more. In the second problem she poses to her students, she wonders what would happen if they combined two different pieces of paper? How long would the label need to be for two pieces of art paper placed side by side if one measured 19 cubes and the other one was 21 cubes long?

While students' strategies for solving 19 + 21 are not necessarily tied to the blueprint and cubes, Hildy refers them to these tools because she is using them as models to represent student thinking. As three students share their different strategies, she models the action of their thinking with the cubes and also by drawing jumps above the blueprint.

In a minilesson several days later, Hildy has taken down the blueprint and represents student thinking on an open number line. Although the cube line is still available for student use, Hildy no longer refers them to it. She is pushing students toward abstraction; they are now beginning to work with their own internalized number relations. Now Hildy uses the open number line as a tool to represent *computational strategies*. Eventually, over the course of many minilessons (beyond the scope of this CD-ROM), the number line will become a mental mathematical model—a tool for students to think with.

FACILITATION TIP 14

Once participants have reflected on the questions on this page, you might ask them to email their responses to you. Their responses will help you plan the kind of work you will need to do to help them understand how Hildy uses context as a didactical tool to support mathematizing. After participants have completed working on the entire folder "The Role of Context," give them their work from *Hildy's Use of Context* and ask them to write a one-page paper responding to their initial observations. Has their understanding of Hildy's use of context changed? If so, how? Ask them to provide specific examples from the video clips that helped them understand the role of context more deeply.

TECH TIP 1

The scroll bar with the green slider will enable participants to move quickly past one part they do not wish to clip to another that they want to move through more slowly. To paste, click on the clipboard icon, then click on arrow. To delete, click on trash can.

Selecting Moments as Evidence

On this page you will find five video clips. You can select moments by moving the green and red sliders. You can paste moments by clicking on the downward arrow. The trash can allows you to delete if you change your mind.

Try to find several moments in those video clips where the measurement context was helpful for the students. Describe each moment and why at that moment the context is helpful.

Was measurement a good context for the development of the open number line?

Participants can use the digital library as a resource to find and select specific moments on the CD-ROM in which the context was helpful to the students. These might include how the context

- supports and develops students'
  - counting strategies. Alexander and Haley's counting strategy evolves into a model of unitizing (Clip 6);
  - understanding of the big ideas underlying measurement (Clip 10);
- helps students (Clip 11)
  - connect their measurements to the number line they are building (example: "How long was the short side of the blue paper? Where should I put 10?" [on their blueprint]);
  - build a system of number relationships (e.g., the measurement 22 can be thought of as 2 tens and 2 more units [cubes]);
  - develop the idea that measurements are precise. When three different measurements are given for the long side of the chart paper, the class checks the measurement by placing the chart paper below the cube string;
  - begin to work with the idea of number space; when Hildy asks, "What if you just had this strip? Where would 66 go?" students change their strategies for placing numbers on the blueprint:
    — Haley demonstrates the relationship of 66 to 46 and 84 by modeling number space with his hands;
    — Lucero and Emily use the number line as a tool to think with (e.g., when they take two jumps of ten from 46 to place 66 on the blueprint these jumps are no longer connected to the cube string; ten is used as a tool);
  - realize that a measurement includes the entire distance. When Hildy asks them, "Where should I mark 66? At the cube or at the edge of the cube?" students are unsure. She brings them back to the context, "My father's friend is measuring the sentence strip where should he measure . . . how can he measure a strip that's 66 cubes long?" and they realize that a measurement includes the entire distance and therefore must go to the edge of the cube;
  - understand that the distance between two numbers is constant (e.g., when Janet realizes that the relationship of 66 to 60 is 6 cubes [units] whether you go forward or backward on the number line);

TECH TIP 2

For technical information on how to use the Paper tool, refer to the Help file under the Tools menu.

FACILITATION TIP 15

There are *Selecting Moment as Evidence* pages throughout this CD-ROM. Although there have been many pages where participants have made comments or have written answers to questions about what they have seen, these *Selecting Moments as Evidence* pages are different. These pages provide opportunities for participants to find and use *evidence* from the clips to substantiate their statements.

The *Selecting Moments as Evidence* pages take participants back over what they may view as old ground, since they have spent much time viewing "The Class at Work" and Hildy's *Building the Blueprint*. They may be surprised as they revisit the clips that their own original impressions have changed in the course of their work with the CD-ROM and that they now bring a different understanding. As they work with *Selecting Moments as Evidence,* their perspective may continue to shift because now they can view the video clips with a highly focused lens and explore in-depth areas (e.g., the role of context, the role of the teacher, and building a community) that they may have only touched on peripherally before.

FACILITATION TIP 16

In the previous sections, participants thought about the structure of the measurement context, where investigation began, how it unfolded, and how it helped the students. In order to understand Hildy's use of context as a didactical tool, participants now need to explore how she builds this investigation through carefully selecting which materials are to be used, when and how they are introduced, and when and if they are removed. The following questions will help participants think about these decisions:

- What materials does Hildy provide for her students? When does she provide them?
- How does Hildy's choice of two colors of multilinks support and develop students' counting strategies?
- At what point in the investigation does she change the ways in which these materials are used? Why does she do this? How does this change support students' mathematical development?

- grounds students' addition strategies in measurement (in Clip 52 when two pieces of paper, measuring 19 and 21 cubes, are joined, how long does the label need to be?).

Building a Learning Environment

Hildy built a learning environment for her students by making several decisions before and during the investigation. For example:

- *She provides multilink cubes to measure the length of the sides of various papers.*
- *She made available at each table bins with two colors of multilink cubes.*
- *At the board she provides a string of green and white cubes in groups of five.*

How did these and other decisions support students' growth and development?

FACILITATION TIP 17

Hildy's learning environment is created in conjunction with her context and supports its development. To participants, Hildy's teaching choices may be puzzling. They may wonder why she

- places two colors of cubes in the bins (why not three or four different colors?);
- does not give the students cubes grouped in fives to *help* them;
- does not give the students rulers so they can get a "real measurement" and why they do not work with these "real measurements" to build a blueprint;
- does not provide them with more information or *hints* before they go off to investigate to nudge them in the direction she would like them to go (e.g., so that everybody has grouped their multilinks in fives or tens);
- puts cubes on a string so that they can move and groups these in fives (why not group them in tens?);
- puts the cubes and paper strip where students *cannot* manipulate them to demonstrate their counting strategies to other students;
- marks different measurements in different ways on the blueprint (e.g., why are the numerals 10 and 20 written bigger than 22?);
- does not mark all the counting strategies on the blueprint (why not mark the *fives* when someone skip counts by five?).

It will be surprising to some participants that, in designing her learning environment, Hildy gave careful thought not only to her choice of manipulatives, but also to how they would be used and when they would be taken away. This is an important idea for participants to think about in this section—manipulatives need to be carefully selected to support the mathematical ideas that are being developed. For many participants the idea that *inappropriate manipulatives can actually hinder and confound student thinking* will be surprising.

As you listen to participants as they work in this section, some of their comments can be very useful to bring up in a whole-group discussion. For example, the questions posed by one participant, "Why micromanage the students' choice of cubes? Why not put out a bin with a lot of different colors and let students decide how many colors they want to use in their measurements? When Hildy controlled their use of cubes, in some ways, she limited their strategies," became the basis of rich conversation about how Hildy used these materials to support the development of the context.

In this section, participants think about what it means to build a *learning environment* and how this environment supports students' mathematical growth and development. For some participants who think about a learning environment as merely the physical layout of a classroom: seating arrangements, learning centers, storage areas that are accessible to students, traffic patterns, and so on, this will be a challenge.

Hildy's vision of a learning environment goes way beyond the physical layout of her classroom, and profoundly influences her teaching. Critical in her learning environment are the *kinds of learning experiences students have* in mathematics, the investigation she structures, and how she employs *context* to support development.

Hildy uses a math workshop model. Here all the elements (investigation, whole-group share, and minilesson) cohere. She starts with students' explorations and inventions and works to scaffold and challenge student thinking. Because she understands the landscape of learning for mathematical development, she can predict possible solutions for the measurement context she poses, and anticipate possible difficulties students might have as they build the blueprint. She understands that learners need to be supported, but also recognizes how important it is to stretch student thinking and challenge misconceptions. For these reasons, her teaching plan could never be haphazard; it is meticulously planned to the smallest detail, including a careful choice of materials.

It is in this light that Hildy's choice of materials becomes critical. Because she envisions the students at each juncture of the investigation, and is clear about the mathematical goals of the sequence of activities, she knows exactly what materials to provide and when to use them.

Hildy sets out two colors of cubes in bins for students as they begin the investigation; how they use them is up to them. While she knows some students may count by ones and may not organize their cubes in groups, she will push their thinking in the whole-class share with the structured cube line she has created. She knows this cube line, organized in alternating groups of five, is a critical tool (along with the blank paper strip) for working with number relations and building the open number line model.

Choosing the Numbers

◉ *Hildy also precut all the art paper to ensure that certain numbers would come up; and, as the children discuss and place their measurements on the line, she continues choosing the numbers carefully. Look back over the whole-group discussion. Note Hildy's choice of numbers. How does this decision making affect children's learning?*

Using Context to Scaffold Learning

◉ *Hildy chooses these numbers to scaffold children's thinking. What is the effect of her choice of numbers on children's thinking?*

In addition to planning which manipulatives to use in the measurement sequence, Hildy also selects the numbers she uses for the blueprint (10, 14, 22, 35, 46, 84, 66, 80, and 60). To get these numbers as measurements, she precuts the art paper, a planning detail that sometimes startles participants!

Because Hildy is thinking about developing the open number line by building number relationships, she initially works with numbers that are close in range (10, 14, 22, 35, and 46). These numbers can be used to build a blueprint in which ten is structurally important and can be highlighted as a landmark.

In order to build a system of landmarks, children need to have ten as a foundation, hence Hildy's starting place. Ten is supported by the cubes arranged in alternating groups of five (white and green). Since the cubes on the string are organized in groups of five, Hildy anticipates that as students try to place measurements on the blueprint, they will work with these important structures. For this reason she focuses the beginning of the math congress on ten and a number close to ten (14). It is interesting to note how she also highlights the structural importance of ten as she represents student strategies on the blueprint. She does not represent *all* their thinking (e.g., even though the first strategy begins with five and five makes ten), Hildy marks ten, and builds a reference system related to groups of ten.

FACILITATION TIP 18

As you facilitate a discussion, you may want to let participants consider *why* Hildy chooses groups of five (as opposed to ten) on her cube line. Some ideas they will begin to consider are that

- five is an important landmark (i.e., its special relationship to ten);
- in contrast to ten, five is a unit that can be easily subitized by the students to help them skip count;
- two fives can be combined to make ten and can be used to help students skip count by ten;
- using groups of five helps students decompose numbers in ways that support ten as a landmark and five as a secondary landmark (e.g., 88 is made up of 80 and 8; the 8 leftover can be thought of as a group of five and three ones).

As participants think about the organizational structure of the cube string, they will also reflect on how Hildy uses it as a tool not only to support and develop students' strategies, but also to build and model important number relationships.

10 14   20 22   30 35

FACILITATION TIP 19

➤ To help participants develop a deep understanding of the role of context in mathematizing, you need to examine two important ways Hildy uses the context to support development. These are her use of (1) potentially realized suggestions; and (2) constraints. While these two didactical tools are implicit in how Hildy uses context, they need to be made explicit in discussions with participants.

Using potentially realized suggestions. As she builds her learning environment, Hildy makes specific choices that have the potential to develop and shift student strategies. These include

Hildy's Choices	*Potentially Realized Suggestions*
Putting out only two colors of multilinks in bins	Students can group these in alternating patterns of five or ten and skip count
Cutting the paper to specific lengths, which includes a range of measurements	Students' strategies for measuring shorter lengths will change for longer ones (e.g., moving from counting by ones to grouping strategies to help them keep track)
Having a cube string grouped in fives	Students use five to skip count to place a measurement
Placing the blueprint below the cube string	Students use the organization of the cube string to place numbers on the blueprint (e.g., five and ten become important tools)
Using paper lengths that build from ten	Students use ten as a reference point (e.g., 14 is ten and 4 more; 22 is two tens and 2 more)
Building a system of tens on the blueprint (e.g., writing the tens larger than other numbers)	Students use ten as a tool to place numbers and to move from one number to another
Changing how the number line is built (e.g., initially Hildy builds the number line with increasing magnitudes) so that students must now consider where to place one measurement in relationship to two others (e.g., where does 66 go in relationship to 46 and 84?)	Students think of number as distance on a number line and develop number space
Removing the blueprint, but keeping the cube string	Students use the number line as a tool to think with

Using constraints. Constraints are another important didactical tool employed by Hildy. In *Building the Blueprint,* Hildy uses two constraints that shift student strategies: (1) Hildy places a cube string organized in fives where students cannot manipulate it. Because they cannot manipulate the cubes to place their measurements, students use the structure of the cube string to count. This moves them away from counting by ones to using different skip counting strategies. (2) Hildy does not use the actual paper they have measured (except to find the measurement for the chart paper), which forces students to mentally construct the measurements in relationship to each other. Both of these constraints are catalysts that move students beyond the situation and help them to generalize important number relationships critical for using the number line model as a tool to think with.

Student strategies reflect this support. As they attempt to explain where a measurement goes, student strategies are fairly consistent. They either

- **skip count by fives to the nearest ten or five** (e.g., Emily consistently skip counts by fives to figure out where numbers would go on the blueprint);
- **work with ten as a unit** (e.g., Josué's initial strategy to place 35, "Take three jumps of 10," which changes when Hildy asks him how she would do that) or **with the subgroups of five within the ten** (e.g., Josué's explanation after Hildy questions him, "Start at 10 [which has already been clearly notated on the paper strip] 5 + 5 makes 10, that's 20, 5 + 5 makes 10, that's 30").

Students work with the strategies listed above to get to landmark numbers and count on the extra cubes by ones or in a group (e.g., Amirah says for the placement of 22, "10 + 10 = 20 . . . and then put 2 more").

When she gets to 46 and its placement on the blueprint, Hildy shifts gears by refocusing the discussion on the measurement context. It is not an arbitrary decision—she knows that she will soon be removing some of the scaffolds she has built with the children. But before she does that and pushes students to work with more complex number relationships, Hildy takes time discussing the different measurements students have obtained for the chart paper. She holds a piece of chart paper against the cube line for everyone to see how long it is. Why demonstrate that at this point in the congress? In doing this, she reinforces an important point—measurements have to be precise, an idea she returns to with three questions: *Where is 66, on the cube or off the cube? Where is 80 in relationship to 84? And where is 60 in relationship to 66?* Hildy returns to the measurement context at this point because she knows that these questions pose enormous difficulty for students. If left unresolved they can hinder students' understanding of big ideas in addition and subtraction. One way to resolve *Where is 66, on the cube or off the cube?* is to reconnect the discussion to the measurement context.

Hildy's number choice takes a huge jump next, going from 46 to 84. Here she removes one of the scaffolds she has been using to support student thinking. For the first time in building the blueprint, there is a distance between measurements that is much greater than ten (i.e., each measurement added to the blueprint differed from the previous one by a little more than ten). Now there are enough measurements on the blueprint for students to begin working with those numbers instead of going back to zero (the beginning of the string) to figure out where a measurement goes. Such changes in student strategies, however, are not necessarily immediate. Take Sarah's strategy: she skip counts by fives from the beginning of the cube line to 80, adding on 4 more.

It is here that Hildy removes another crucial support that she has built. Thus far students have been building the number line in a forward direction; now they are going back (from 84 to 66). To find 66, however, they do not need to work backward, (although they can if they choose to), but also can make connections to 46, which is two tens away. (It is important to remember that a lot of time was spent to emphasize the placement of 46.)

FACILITATION TIP 20

The questions listed below can be helpful to keep in mind as you work with participants in the folder, "The Role of Context," and as they think about Hildy's number choice:

- Why might Hildy have picked the numbers she uses? How are these numbers connected?
- How does she structure her choice of number?
- Where does she begin? Why do you think she begins there?
- How does she record the numbers on the recording strip? Why is she doing this?
- At what point does she change her use of the number? Why does she do this?
- How does her choice of number affect student thinking?
- How does her choice of numbers scaffold student thinking?
- Do student strategies for placing numbers on the blueprint change? If so, when?

Prompted by Hildy's number choice and the careful way she has been working with number relationships, student thinking does shift in two powerful ways. Lucero and Emily work with the idea of starting at 46 and *using jumps of ten* (i.e., Emily moves from 46 to 56 to 66, each time *jumping* a group of 10). These jumps of ten, however, no longer coincide with the cube line; now the jump of ten is not visually connected to 5 + 5, but has become 4 + 5 + 1. Both girls struggle with this physical representation of their mental actions on the cube line. While she puzzles about how to represent her thinking with the cube line, it is clear that Emily's mental actions *no longer* need this support.

This is a big moment because students' mental actions now include thinking *with ten (in jumps of 10)*. Their mental actions can be represented as jumps on a number line (which Hildy uses as she represents Emily's thinking). This is an important step in the development of the open number line as a model to represent student thinking.

Tied to Hildy's choice of number is a dramatic event reflected in Shannon's strategy. Shannon's description of where 66 would go includes both 50 and 70 as reference points in understanding the relationship of 46 to 84. His hand motions parallel jumps on a number line. He moves backward from 84 to 70 and forward from 46 to 50 saying that "60 has to go in the middle." He is physically modeling number space and moving within that space in any direction. He is beginning to use the model as a tool to think with.

As Hildy's use of number unfolds, it is clear that she is using the measurements of the art paper to do more than build a system of number relationships. She is creating a model for number that students can work with. This model is open and can be used to represent thinking.

The Role of Context

How does Hildy's choice of context foster mathematical development?
 Hildy often reminds the children about the context as a way to enable them to imagine concretely what they are doing. Use the green and red sliders to find footage

where Hildy brings the children back to the measurement context. You can paste moments by clicking on the downward arrow. The trash can allows you to delete if you change your mind. How has the representation of the students' thinking evolved?

There are a number of moments where Hildy uses the context to help students realize what they are doing. Participants' selections here might include how Hildy

- connects the measurements they have taken to the blueprint they are building (How long is the short side of the blue paper? Where should I mark 10 on this blueprint?);
- lines the actual chart paper up against the cube string to help students obtain a precise measurement when there is disagreement about a measurement;
- returns to the art show context and cutting a label for the art paper that is 66 cubes long to help students realize that this measurement goes to the edge of the cube;
- grounds their addition strategies in the measurement context by connecting
    - an addition problem to the context (When two pieces of paper measuring 19 and 21 cubes, respectively, are joined, how long does the label need to be?);
    - students' solutions to this addition problem to the context so that their strategies can be physically modeled for the other students (e.g., Hildy models Amirah's strategy (21 + 10 + 9) by asking her where 21 would be on their blueprint and, after recording 21 between 20 and 22, shows the jumps in Amirah's strategy with the cube string and blueprint).

Backburner: A Moment for Further Reflections

◉ *This is the last page in the folder, "The Role of Context." However, you may have other questions on this topic that you would like to investigate. Go to the TOOLS menu above and add them to your backburner notes.*

See *Backburner* notes, page 8.

THE ROLE OF THE TEACHER

Introduction

◉ *In this folder, about the role of the teacher, you can investigate how Hildy supports development, questions, facilitates dialogue, and provides "think time."*

Supporting Development

◉ *There are many moments where Hildy does subtle things that support mathematical development. Use the green and red slider to select some moments that you notice. You can paste these moments by clicking on the downward arrow. The trash can allows you to delete if you change your mind. Justify your choices.*

In this folder, participants consider ways in which Hildy, as a teacher, supports mathematical development. They look for evidence in the CD-ROM (there are clips from *Children at Work, Building the Blueprint,* and "The Next Day: Mental Math," and "Several Days Later: Mental Math") and document their findings by clipping specific moments that support their ideas.

For facilitators, this section is a window into participant thinking. If you have used the CD-ROM for an extended period of time, it is also a way to assess not only their understanding of the role and use of context in mathematical development, but also the ways in which a teacher uses pedagogy to support learning. If participants have worked with the folder "Growth and Development," they can begin to think about *how* the changes they saw in some students were supported and affected by Hildy's decisions as a teacher. Although the clips participants choose as evidence will vary, there are several major areas that will inevitably come to the surface.

1. **There are many mathematical choices that affect students' development. Hildy**
   - carefully chooses numbers to support and scaffold student thinking;
   - structures the discussion in building the blueprint around landmarks and progressively builds on these relationships;
   - alters student thinking by changing the rhythm of the discussion with three important questions: (1) *Where is 66?* (i.e., in relationship to 46 and 84); (2) *Where should I mark 66: at the cube or at the edge of cube?*; and (3) *Where is 80?* (in relationship to 84);
   - spends a good deal of time before this discussion placing 46 on the blueprint, which is a pivotal number in the following discussion (e.g., *Where is 66?* etc.);
   - removes some of the scaffolds to help students think about the placement of measurements without the visual support offered in the cube line (e.g., to the question *Where is 66?* she asks Emily, "Could you figure out where 66 would go without using the cubes?" Emily, who has been going back to the beginning of the cube line each time to prove where a number goes, begins to think of specific number relations not connected to the organization of the cube line in fives (e.g., using ten as a tool to move from 46 to 56 to 66). Hildy's question moves Emily on as an individual, but also sets the stage for the entire class to begin working with mental math and the open number line as a model;
   - chooses numbers in her second minilesson (43 + 20) that are connected to the discussion around where to place 66 (two student strategies worked with adding two tens [20] to 46).

2. **Hildy creates a rich context to**
   - initiate students' explorations in measurement and give them *thinking space* to make sense of the problem;
   - use as a didactical tool to develop the open number line model;
   - help students make meaning (e.g., she brings them back to the context to support their thinking: (*Where is 66: on the cube or off the cube?* so that within the measurement context, students' initial misunderstandings are resolved);
   - support student work in addition (e.g., 19 + 21 is introduced to the children with the story of putting two pieces of art paper together and wondering how long the label needs to be).

3. **There are pedagogical choices that support students' growth and development. Hildy uses**
   - a quiet thumb signal when they are ready, which helps students who need more time to think, but at the same time validates those who have finished quickly;
   - think time to give students enough time to formulate their answers;
   - pair talk and student paraphrasing at crucial junctures not only to emphasize the importance of students' listening to each other, but also to highlight important mathematical ideas;

24 Journey 2

- open-ended questions that probe student thinking and also help them express their ideas more clearly (e.g., *And where should I mark that? How do you know that?*).

4. **Hildy's choice of materials also supports students' mathematical development. She**

    - gives student pairs one clipboard and has them share it. This *sharing* requires that students talk to each other and reach consensus when in disagreement;
    - gives them a premade student recording sheet;
    - puts out two colors of cubes in the bins knowing that this is connected to the two-color cube line grouped in fives that will be used in building the blueprint;
    - precuts the paper students measure to ensure that certain measurements come up in their whole-group discussion;
    - uses the cube line to support the creation of the blueprint;
    - initially models student strategies with the materials; this evolves from modeling their thinking with the paper strip and cube string to using the open number line.

5. **Hildy also supports students' mathematical development by supporting their social needs within the classroom community. She models the behavior she expects her students to exhibit. She models**

    - what good listeners do by the ways she attends to each student's ideas;
    - respectful behavior toward each student by validating and listening to their ideas;
    - that learning is hard and requires effort (note how she puzzles over student answers, trying to make sense of their ideas);
    - how to ask questions—that, if sense making is important to learning, then being a student entails asking clarifying questions when confused;
    - that disagreements are not personal attacks and defending one's ideas is essential to being a member of a *community of discourse*;
    - that she is a learner too, and as such can make mistakes and change her thinking. Sample dialogue is provided in Appendix C, page 45.

FACILITATION TIP 21

Part of participants' work with "The Role of the Teacher" includes reflecting on Hildy's use of materials (why did Hildy pick certain materials, were they effective, etc.). The *student recording sheet* (located to the right on each page in *Children at Work*), which offers insights into each group's thinking, might prove useful in a conversation about Hildy's choice of materials. Questions to use in this discussion are

- How is the recording sheet structured? Why might Hildy have structured it like this?
- How do students use this sheet? Does the structure help or hinder the students? What evidence do you have to support what you are saying?
- Do their recordings reflect their strategies (is there a connection between the struggles you see children having as they measure and the ways in which they record their measurements)?
- Do their recordings give you any additional insights into their thinking?
- What might be the benefits to structuring or not structuring a recording sheet?

Maximizing a Moment

◉ *At a certain moment, Hildy asks the students whether she needs to put the mark at the cube or at the edge of the cube. What arguments do the students have for each point of view? Why is this discussion important? How is her question related to the context and to measurement?*

The discussion Hildy has around the question Where does 66 go? is an important one mathematically. Student confusions on this contribute to difficulties in subtraction. What do I go back? Where do I start counting? What am I counting, the space or the cube? (This actually happens with Janet, who counts back three when asked where is 80 in relationship to 84.)

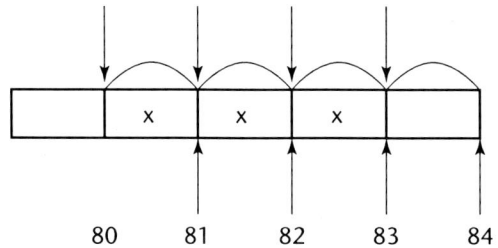

Go back four spaces from 84?
Count four cubes back starting with 84?
Count back three cubes (and not include the 84th cube in your count)?
Count back four spaces from the end of 84?
Where you start counting and what you are counting changes your answer.

Initially in this discussion, students argue that 66 is on the cube, not at the edge of the cube. The discussion is critical because it brings to the surface the difference between ordinal and cardinal numbers and highlights an idea that is often confusing to students. Where the mark goes does not matter if you are thinking of ordinal numbers (the 66th cube can be on the cube or in the middle of the cube); it matters greatly in terms of cardinality, and particularly when measurement is involved. Confusions around this idea surface in subtraction when students grapple with *What am I taking away (or going back)?*

At this point in building the blueprint, students struggle with this issue, giving various opinions (the middle of the cube, on the cube) about where 66 should go. One student, Lucero, believes 66 is at the end of the cube, but only because Hildy has marked it as such. Her explanation does not touch on *why the mark goes there*. Some students express the idea that the edge of the cube is too close to the next number, 67, or is that number. Part of Shannon's confusion as he argues for placing 66 on the cube is connected to Hildy's tool for marking numbers; he demonstrates with the pointer (which has a girth near the size of a cube), that if they mark the cube where the pointer ends, "it is the next cube."

Once these ideas are on the table, Hildy brings students back to the measurement context. In light of building the blueprint, it becomes apparent to students that 66 means the entire distance (the measurement) and as such, has to go to the end of the cube. If it is shorter than this, the label for the art show will not match the length of the art paper. Sixty-six must include *all* of the 66 cubes.

Posing Important Questions

◉ *Find other moments where Hildy asks important, challenging questions. Select and paste some moments and justify your choices.*

In this CD-ROM, participants are asked to *find moments* where Hildy asks important or challenging questions. There are many examples that participants can select to illustrate Hildy's use of questions; these include using questions to

FACILITATION TIP 22

Participants will have no trouble identifying and clipping Hildy's questions; there are numerous examples of these throughout the CD-ROM. How they justify their choices, what they offer as examples of important or challenging questions, will give you key insights into their thinking about the role questioning plays in teaching.

- **affect change** (example: Clip 11, when Emily goes back to the beginning of the cube string to place 66, Hildy says, "So you were using the blocks again. What if you just had the strip at this point, could you figure it out just using the strip?" Emily uses the blueprint to place 66);
- **cause puzzlement** (example: Clip 11, when Hildy asks, "Should I mark 66 on the cube or at the edge of the cube?" students spend quite a while puzzling over this question);
- **support student reflection by encouraging them to talk about their strategies** (example: Clip 11, when Josué shares his strategy for placing 35, Hildy asks him, "Can you tell me how I'm going to do that—three jumps of ten?" Josué explains how he uses two groups of five to count by ten ("5 + 5 makes 10, that's 20, 5 + 5 makes 10, that's 30");

- **promote classroom dialogue** (example: Clip 11, Hildy asks, "Where would 66 go?" and has students talk to each other about this before having a whole-group discussion);
- **encourage students to comment on ideas presented to the group** (example: Clip 11, after Amirah places 22 on the blueprint, Hildy asks, "How can I be sure that's 22? Does somebody else have another way they can explain it?");
- **deepen students' understanding** (example: Clip 11, "If we think about measuring those strips, if we went to the middle of this cube and called it 66, would we get a strip that was 66 cubes long?" When she brings back the measurement context, Haley changes his original idea (that 66 was on the cube) and says, "We get the 66 at the end of it").

Hildy also uses questions to help students clarify their strategies, which not only supports individual learners in the articulation of their thinking, but also helps other students who may be struggling to understand a given strategy. (A wonderful example of Hildy using questioning in this way is during the beginning of *Building the Blueprint*. Amirah's explanation for placing 22 on the blueprint is rapid fire and unclear, "Those two are ten and then you add two more." Hildy slows Amirah down by puzzling over what she is saying; her careful questioning helps Amirah clearly articulate how she knew where 22 would go.)

ACTIVITY 1 The Role of Questioning

Before setting off to work with the CD-ROM, you might want to explore participants' ideas on the role of questioning. Toward this end, you might ask them to write a paragraph or two on how they use questioning in their own teaching (if participants are preservice teachers, you might ask, What role does questioning play in teaching?). The information from this assignment can help you plan your work with participants. If their understanding of the use of questioning is shallow ("Teachers use questions to find answers." "Teachers pose questions to help students learn."), allow ample time for digging into their ideas by juxtaposing them with how Hildy uses questioning on the CD-ROM.

At the end of working with *Posing Important Questions*, you might also ask participants to revisit their initial ideas and reflect on how (and whether) their ideas on the role of questioning changed as they worked with the CD-ROM clips. In this assignment, you could have participants use the Paper tool (see TOOLS on the menu bar) and hypertext specific clips from the CD-ROM that might have reshaped or supported their thinking about the ways in which questions can be used to facilitate learning.

Facilitating Dialogue

Let's focus on Hildy's actions during the math workshop. For instance, why might Hildy be pushing children to paraphrase and explain what other children are doing?

Pair Talk

At other moments during the math workshop, you see Hildy asking the children to have pair talk. In the first video clip below you can see one of those moments. After looking at it, try to find several more of moments of pair talks. How does pair talk support development? Are there other moments where you would like to have pair talk too? Justify your choices.

Hildy knows that the construction of meaning is not merely an individual activity, but involves the entire classroom community. Social interaction, the sharing of ideas through conversation, engenders further thinking. For this reason, in her classroom, learners are expected to communicate their thinking.

Hildy's role as a teacher is twofold: she needs to support individual students' growth and development while working within the classroom community to establish mathematical ideas. In order to create a community of discourse, Hildy uses several important pedagogical tools: think time, pair talk, and student paraphrasing of each other's ideas. She uses these tools at different times and for different purposes.

Throughout the discussion in *Building the Blueprint* and the ensuing mini-lessons (in "The Next Day: Mental Math" and "Several Days Later: Mental Math")

Hildy gives her students think time, asking students to use a thumb signal to indicate when they are ready. By allowing them time to think, she exhibits a respect for student intelligence and also signals that making sense is not an immediate event, but takes time and serious work. These two tools help her to pace her conversation and to show that getting an answer quickly is not the point, but thinking *about your answer* is.

Other pedagogical tools like pair talk and paraphrasing are used at critical moments in the discussion: (1) to emphasize an important mathematical idea; (2) to highlight a student's observation that has the potential to shift student thinking in profound ways; and (3) to work with student disequilibrium. Paraphrasing is also a way to empower students, to let them know that they are responsible for listening to each learner and that making sense of other people's thinking is an important part of what it means to be a student. The work is not just about one's own thinking, but also about respecting the thoughts and ideas of others.

FACILITATION TIP 23

It is important in your work with participants that you explore the ways in which they use pair talk so that you can compare these uses with Hildy's. Some participants will recognize that Hildy's use of pair talk is very different from their own but not be able to distinguish how it is different. It is critical here to help them delineate how Hildy uses pair talk as a powerful pedagogical tool to work with student disequilibrium, to shape and support development, and to highlight important mathematical ideas presented to the classroom community. It is also important to not necessarily see Hildy as an exemplar, but instead to encourage participants to look for other moments where the technique might have been helpful, too. (A sample dialogue is provided for you in Appendix C, Dialogue Box E, page 42)

These pedagogical tools reflect Hildy's views about teaching and learning. When used wisely, they are effective means of supporting the idea that learners exist in a community, that ideas are not only for the teacher, but for everyone. In this classroom, the dialogue is not just between teacher and student, but also includes student-to-student interactions. This is an important structure that empowers students not only to challenge each other's ideas, but even those of the teacher!

Providing Think Time

◉ *Observe the amount of time Hildy gives the children to think in the following episodes. She also has the children use thumbs to show when they are ready. How does think time support development?*

The discussion in *Building the Blueprint* can be seen as a model of how Hildy uses pedagogy to support student growth and development. First and foremost, it is clear that she values students as learners and respects their ideas. Through her behavior she also models a belief system that recognizes that learning is difficult and that *all* students are capable of thinking. Because thinking requires serious effort and does not necessarily happen quickly, Hildy gives her students ample time to ponder and respond to her questions. To provide this learning space, she uses two pedagogical tools: think time and a quiet thumb signal. For those children who find an answer more quickly, she lets them communicate this, but they do it in a quiet way (no hand waving or anxious body movements to signal completion of a task or an answer to a question) so that they are not interfering with the thinking of other students. It is clear from Hildy's actions that she respects and values individual differences.

Backburner: A Moment for Further Reflections

◉ *This is the last page in the folder "The Role of the Teacher." However, you may have other questions on this topic that you would like to investigate. Go to the TOOLS menu above and add them to your backburner notes.*

See *Backburner* notes on page 8.

DEVELOPING A COMMUNITY

Introduction

In this folder, about developing a community, you can investigate what Hildy does to establish a challenging, but safe, environment for learning.

What is a community of learners? How does a teacher create and sustain such a community? These are not new questions and many participants may have very specific ideas about how teachers create communities of learning in their classrooms. Their own ideas may center around things like creating a learning environment where students are respectful toward each other (and the teacher), where there is accountable talk and listening, and where students can work independently or in groups and *stay on task*.

In Hildy's classroom, learners are empowered by her expectations, which are high. She expects them not only to think and to struggle in the process of learning (learning is hard work), but also to sustain their efforts and persevere.

It is clear from her behavior (see "The Role of the Teacher" for a more in-depth analysis of specific pedagogical tools that affect student growth and development) that she values their thinking and is interested in what they have to say. Because she respects their ability to think, she gives students a real problem to solve, and lets them find their own solutions. She uses their ideas and strategies as the basis for deepening their thinking in her math congress. Here her questions probe their thinking and help them clarify their ideas. She lets them struggle both to understand and to express that understanding.

What she expects from each individual learner, she expects from the class as a whole. In this community of learners, there is not only respect for the power of ideas, but also a great emphasis is placed on being able to clearly communicate one's thinking to the class. Ideas are proposed to the group and students are expected to understand, question, or disagree. These expectations create a community of discourse, a fundamental feature of Hildy's classroom.

No one is below these expectations—this is the foundation from which Hildy builds her classroom community. Students are expected to think, to have ideas, to be able to share them with their peers and to be able to defend them if questioned. That students have internalized these expectations is clear from the ways in which they treat each other. During the investigation, children worked in pairs. Their work was thoughtful and purposeful. While there was real collaboration between the students in each pair, there were also disagreements. These were resolved by the students, not by the teacher. The autonomy reflected in student behavior is fostered by Hildy in another important way. Over the course of the CD-ROM, we see students *challenging* Hildy. In one instance during the first minilesson, Amirah asks Hildy to write down her strategy on the board and gets a little annoyed when Hildy does not understand her strategy. Hildy questions Amirah, but asks her to explain her strategy in terms of the cube line and the paper string. Amirah holds her ground, saying, "I'm not doing [using] that," and goes on to explain her strategy.

In another instance, Emily challenges Hildy when she places jumps above the paper strip to model her strategy. Emily says, "It wouldn't be right there because then you wouldn't have room for 70." We see Hildy thinking about Emily's idea and agreeing with her. Hildy then changes her notation of jumps to reflect Emily's suggestion. Students can challenge Hildy because she too is a member of the classroom community; *they hold her to the same expectations she holds them*. Hildy is a member of this community and not above being held accountable for her ideas!

Evidence of a Community

Hildy and her students form a community of learners. Use the green and red sliders to select footage where you see evidence that there is a community in this classroom.

You can paste moments by clicking on the downward arrow. The trash can allows you to delete if you change your mind. Justify your thinking.

After participants have clipped their moments, it is important that you examine some of them together. The intent of this activity is to examine what a community is and how it functions—and to look for evidence of these characteristics in the footage. It is *not* to look at the role of the teacher in establishing a community. If participants focus on the teacher and what she is doing too early, they only look for what the teacher is doing to establish *their* notion of community and it is just this—their notions of community—that you may need to challenge first.

ACTIVITY 1 Clipping Evidence

The clips on the page *Evidence of a Community* show footage of the investigation in four segments. Participants will find that some clips are more useful than others as they search for supporting evidence.

It is important to allow participants to choose moments without prefacing what they should look for. This will allow you to subsequently examine participants' beliefs about community, what they even think a community is! Do they think community means listening to the teacher unless called on? Or are they able to appreciate the way children question each other and work autonomously without a teacher sitting with them? Do they know the appreciation for ideas and thinking? Do they see Hildy's handling of mistakes, which they examined in the last folder, as evidence of trust?

Building and Supporting a Community

What does Hildy do in this lesson to establish or support this community of learners? Again, select important moments in this lesson where you see evidence of the teacher building and supporting this community. Describe the moment, and describe how Hildy supports this community of learners.

Participants in the previous section, *Evidence of a Community*, selected moments in the video footage that they felt *exemplified* community and justified their selections, and you worked with them to help them appreciate how a community has trust for its members, how dialogue flows, and how supporting and exchanging of ideas is an important characteristic of a functioning community. But how does one develop a classroom community?

To answer this question, participants are now asked to revisit the same footage to think about how Hildy, in her interactions with individuals, small groups, and the entire class, creates and consistently nurtures this community. This is a slightly more difficult task because participants have to contemplate the subtle pedagogical tools Hildy uses as well as interpreting her behaviors as she interacts with children. It is easier for participants to identify the elements that make a community (e.g., "Look at how respectfully the children speak to each other") than to isolate the behaviors and words of a teacher that *make this happen*.

ACTIVITY 2 Mapping the Dialogue Ball

One way to develop participants' understanding of how Hildy uses dialogue to work with mathematical ideas as she simultaneously supports the development of a community of learners is to have them imagine the flow of dialogue in a conversation as a bouncing ball that goes from one person to another to another and back again, as they share ideas and question one another. This image may be helpful to participants as they respond to the questions in this section. Hildy holds the ball in the beginning. To whom does she bounce it? Where does it bounce from there? When does she want the ball? What does she do to keep it from bouncing back to her all the time?

Ask participants to analyze the flow of conversation, the *dialogue ball*, and think about how Hildy

- shifts the conversation;
- what reasons she has for doing this; and
- how and why she involves different learners in the conversation.

After they have had time to examine the CD-ROM clips, have a whole-group conversation, but focus the discussion on

- What kinds of questions does Hildy use to throw the ball to the students?
- How does she ensure that they pass the ball to each other and not immediately back to her?
- Are students autonomous? Can they take the ball without being given it by their teacher?

ACTIVITY 3 Developing a Community

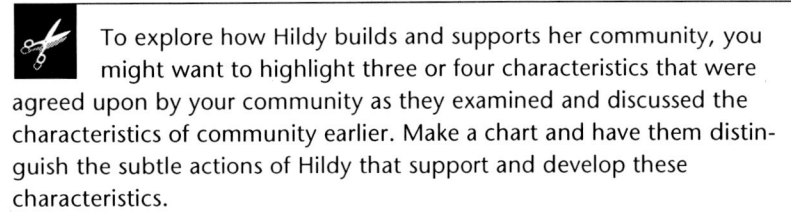

To explore how Hildy builds and supports her community, you might want to highlight three or four characteristics that were agreed upon by your community as they examined and discussed the characteristics of community earlier. Make a chart and have them distinguish the subtle actions of Hildy that support and develop these characteristics.

Some participants may think the children are special, nice children, and that Hildy is just lucky to have such a group. To understand that this sense of community is being *developed* in front of their eyes, they need to examine all the (almost invisible) things Hildy is constantly doing to establish, support, and develop this community. The dialogue pattern is one of the critical pieces and participants have just examined that, but there are others.

Backburner: A Moment for Further Reflection

◉ *This is the last page in the folder, "Developing a Community." However, you may have other questions on this topic that you would like to investigate.*

Go to the TOOLS menu above and add them to your backburner notes.

See *Backburner* notes, page 8.

THE NUMBER LINE

Interview with the Teacher

◉ *In this folder you witness Hildy Martin and Cathy Fosnot talking about the sequence of activities. In particular they talk about the development of mathematical modeling and the emergence of the open number line. On the next page is a representation on a time line of the activities Hildy and Cathy discussed.*

In the interview you hear Hildy Martin as she tells Cathy Fosnot how she has planned the measurement context so as to construct, with the children, a blueprint of the lengths of the art papers they had measured. A building analogy and its vocabulary work well to describe the sequence of activities. Hildy has scaffolded the context with a variety of precut art papers and the two colors of multilink cubes for the children to measure with. Their constructed cube lengths become representations, or models, of their counting strategies and give meaning to the numbers they write on their recording sheets. As she builds the investigation, Hildy adds new supports. In the math congress, the children can use the two-color cube line as a guide, not only for deciding where to place their numbers on the blank paper strip below it, but also for clarifying the relationships between numbers and confusions in moving backward on the strip.

In the first minilesson that follows, "The Next Day: Mental Math," Hildy uses the blueprint to model the students' strategies for solving the addition problems. The jumps they take are represented above the paper strip and supported through her manipulation of the cube line. Several days later, she removes the blueprint and models students' strategies on the open number line. For some students, this model has become a tool to think with.

Seeing and hearing the teacher talk of her goals and planning may be the catalyst that jolts participants into reconsidering their teaching and learning beliefs.

Representing Emergent Modeling

◉ *For each moment on the time line, find clips in the footage that show evidence of the development that Hildy hoped for.*

These last "select moments" give participants another way to review the activities of this CD-ROM to see how they evolved over time into the development of the open number line model. Viewers can come to appreciate the complexities of Hildy's goals and how a deceptively simple activity can form a solid foundation for children's understanding of number.

Backburner: A Moment for Further Reflections

⊚ *This is the last page in the folder, "The Number Line." However, you may have other questions on this topic that you would like to investigate.*
   Go to the TOOLS menu above and add them to your backburner notes.

See *Backburner* notes, page 8.

Epilogue

In this CD-ROM, participants reflected on how context could be used to support the development of the open number line. They studied how this model emerged progressively from the initial measuring of strips of paper (modeling the situation) to the didactical use of it, by Hildy Martin, to represent her children's computation strategies. Eventually, the children even began to use it as a powerful tool to think with. Beginning as a model *of* the situation, it ended as a model *for* thinking.

After working with this CD-ROM, if participants want to explore further the open number line and its use, *The Role of Context: Exploring Ages* or *Minilessons PreK-3* would be good choices. In both of these CD-ROMs the open number line is used by the teacher to represent children's strategies for addition and subtraction. Children are seen also using the number line as a model to think with as they imagine their strategies for computation as leaps to landmark numbers, and/or as taking leaps of tens.

The open number line is an important model that particularly supports the development of computation strategies where one number is kept whole—nonsplitting strategies (e.g., solving 342 + 199 by adding 200 to 342 and then removing 1, or solving 29 + 41 by turning it into 30 + 40). In contrast, the sequence on the CD-ROM *The Role of Context: Taking Inventory* is more powerful for developing splitting strategies, and the standard place value algorithms (carrying and borrowing). A comparison of these contexts might be an interesting topic for discussion.

Appendix A

CHILDREN AT WORK: ANALYSIS OF STRATEGIES

ALEXANDER AND HALEY—CLIP 6

There are three distinct sections to this clip. In the first one, Alexander and Haley measure the long side of the chart paper, the short side, and then the long side opposite the first side they measured. Alexander writes on the clipboard and keeps track verbally of the tens, while Haley does the physical work of touching the cubes and also counting. They are using only one color cube: white.

Haley's strategy for counting evolves over the course of Clip 6. Notice that his right hand acts as his counter as his left hand marks the important intervals. At first these markings indicate groups of five (two fives make a ten, and it is at ten [the second group of five] that he pauses). After Alexander says, "20," Haley's strategy changes and his left hand marks the unit of ten and not each group of five within the ten. The development of his strategy in Clip 6 shows *unitizing*, a big idea on the landscape of learning for early numeracy. In the third part of Clip 6, the evolution of Haley's thinking is clearly manifested in the red cube he uses as a marker for a group of ten. How interesting to note that, after Haley has attached a group of ten white cubes to the larger group lining the chart paper, Alexander begins to count the added white cubes by ones. When Haley puts the red marker on the group, Alexander stops counting by ones and says, "Oh, it's 40." This moment clearly shows how *both* boys are using unitizing as a strategy in obtaining their measurements.

While the big idea the boys are working with is unitizing, there is a discrepancy in their counting that may become obvious to some participants as they watch the clips. In the first part of Clip 6, Haley inaccurately counts the cubes on the long side of the paper. There are subtle things that happen in his hand motion that may or may not contribute to this—his left hand, when he reaches 30, stops marking the ten and physically it is difficult for him to reach all the cubes. As he's counting, he misses a cube; his final count is 45 when the actual length is 46 cubes.

In the second section, Haley pauses at the end of his count of the short side of the paper, and Alexander says, "50." Haley corrects him and says, "51." At no point in the clip do they puzzle over the fact that they now have a longer measurement for the shorter side of the paper (51 for the short; 45 for the long). In the third section, one of their groups of ten is actually 11 cubes long. When the camera moves back to give a full view of their cube line, you can see that they are also one cube short. Thus their second measurement of the long side is 44.

Their student recording sheet offers an interesting trail of their thinking. There is some crossing off; at some point they rethink their initial measurement of 51 cubes for the short side and come up with 29. Why are the boys measuring the opposite side of the paper (a measurement they already have)? How can the short side of the paper have a longer measurement than the long side? Why are they getting two different measurements for the same length? For participants asking these questions, it might be useful to refer to the boys' recording sheet, which offers insights into the evolution of their thinking. Alexander and Haley are grappling with unitizing at the same time they are working through some big ideas underlying measurement. Their way of marking each group of ten with a red cube shows unitizing; it is also a system that helps maintain a more accurate count.

AMIRAH AND JONATHAN—CLIP 9

Amirah and Jonathan each build stacks of green cubes and align them with the paper, adding more cubes as needed. After Jonathan puts his stack of cubes along the paper, Amirah tells him he needs one more cube. The brown cube he adds creates a frame (connecting length and width), which gives them an inaccurate measurement (15 cubes instead of 14). Amirah uses two fingers and counts the cubes by ones (her counting is interesting because, even though she counts by ones, the motion of her fingers indicates a grouping of two).

ANDREW AND SHANNON—CLIP 4

There are three sections in Clip 4 that show the two boys measuring different-colored paper. Throughout Clip 4, the boys use yellow and pink cubes grouped in fives. Although the cubes are grouped in fives, they clearly count by ten. The importance of these two groups of five is manifested in their physical actions as they count.

In the first section, Shannon takes a previously built strip from the long side of the paper to use for the short side. This is an indication that he understands that the measurement of the shorter side is contained in the longer one (see hierarchical inclusion—how numbers nest inside each other—a big idea on the landscape of learning). Andrew, who initially protested Shannon's moving the stack of cubes from the length to the width, also goes to build the opposite side after Shannon counts and Shannon tells him, "We don't need to measure that side, it's the same thing."

Their counting strategy and grouping of cubes (two fives clearly indicated) remains the same throughout Clip 4 and can be seen in Shannon's hand as he counts "10, 20, 30" and touches the second group (the pink cubes) of five that comprise the ten. But their counting is thrown off in the second section.

Here, there is a stack of cubes lined up against the width of the paper. Andrew recognizes that they need another cube. He does not move the stack down and align it with the beginning of the side, adding additional cubes to the top of the stack, but instead adds the cube they need to the beginning of their cube strip before the first group of five. Reading across the cubes, if you are thinking in terms of fives or two fives making a ten, you now see 1 + 10 + 10 + 5. When Shannon begins to count by ones, Andrew stops him and counts by tens. His hand movement here is interesting to note. Andrew ignores the initial yellow cube, and begins by touching two groups of five (pink/yellow) using two fingers of his hand to indicate the two fives that comprise the ten. On the last group of five, he counts another 10 and says, "30," but his two fingers do not touch the groups (they are not there)—so his hand motion is in the air.

Shannon disagrees with him and also uses the yellow/pink grouping to count. What makes Shannon disagree with Andrew becomes apparent after he counts, "10, 20, 30." Shannon indicates the single yellow cube that Andrew omitted, and says, "31." The first ten Shannon counts, however, is really only 1 + 5. There is some disagreement between the two boys and Shannon finally resolves the situation by counting the cubes by ones. Now he counts 26. His statement at the end of the count, "26. How can that be?" is an indication of disequilibrium—he recognizes that he has just counted the same stack of cubes and has gotten two different answers!

ANGELICA AND DIANA—CLIPS 72–75

In the clips of these two girls, it is easy for participants to ignore Diana—she says nothing during the clips. For this reason, it is important for participants to focus carefully on what she is doing. She is building stacks of cubes, using two groups of five (red and green) to make ten. She is handing these stacks of either five or ten to Angelica, who does not use this structure to count.

This is apparent from the beginning of Clip 72. The short side of the paper has been measured and the cubes are clearly grouped in fives (alternating green and red). Angelica is beginning to measure the long side of the paper and grabs a small group of red, but then changes it for a group of five green. Diana hands her a stack of ten (five red and five green), which Angelica connects to the five green. Angelica then reaches for another stack of ten, but when it doesn't fit because the multilinks cannot be joined, she flips the cubes and creates a group of ten (green). This breaks the pattern of fives that is clearly discernible. It is important to note that Angelica counts the cubes by ones even though the beginning of the structure could have been counted by fives or tens.

In Clip 73 both girls are measuring opposite sides of the same paper. This clip visually reflects the differences in their thinking: Diana's grouping shows her consistent use of a repeating pattern of fives, alternating green and red groups; Angelica's grouping is random. A subtle, but clear, indication of Diana's thinking in terms of five is shown as she takes a stack of green cubes out of the bin, counts them, and puts one cube back. There were six; putting back one cube shows that she is carefully building fives and is using this structure to help her count (even though we never hear her count). Angelica's grouping is haphazard (she uses her stack from the first clip, five green, five red, ten green, two red, and adds to this). She continues to ignore the grouping and counts by ones.

ANTHONY AND SARAH—CLIP 3

The students are working on measuring the large yellow paper. While they use two colors in their measurement of the length, these are not grouped. They consistently count the cubes by ones. Their first measurement is 83 and Sarah says. "83!!!" indicating her surprise at this length. They set to work measuring the other side. Sarah realizes in the midst of doing this that they need another cube for the length. She clearly counts on from 83, telling Anthony to change the measurement he has recorded to 84.

EMILY AND JANET—CLIP 10

This clip shows both girls working with ten as a unit. They use one color cube (pink) in their measurement, but it is clear from Emily's actions that ten is an important structural unit for her. Emily counts a group of ten out of the stack of pink cubes, and says, "Here's 10." Janet follows her example and counts out the next group of ten, which

Emily takes. As she holds each group of ten and the leftovers, Emily skip counts, saying, "Here's 10, 20, 22." She recounts each stack of ten by ones, but once again, note that the structure of this count is still around the group of ten. This sometimes eludes participants as they watch the girls count and recount. Comments commonly made are that "the girls keep counting by ones." Other participants focus on the fact that once Emily counts a group of ten, she does not use this length to determine that the other stack is also a group of ten. While this is so, it is important to focus the discussion on the mental action that underlies the physical one. Although Emily does touch and count the *groups of ten* by ones, underlying her physical action is an important mental construction—ten is a landmark number. In her final count of the two groups of ten by ones, Emily takes the two groups and says, "That's 20." She clearly knows that 2 tens makes 20.

FERNANDO AND JOSUÉ—CLIP 2

The two boys are constructing a frame of black cubes around the large yellow paper. Fernando measures the stick he's making with the one that's already built for the width and goes to place it on the opposite side.

LILLY AND MELISSA—CLIP 7

The girls have organized their cubes into groups of ten and use two colors, brown and green. They count their cubes as Lilly says, "by tens and ones!" In the first part of Clip 7, they count the cubes four different times. The clip begins with Melissa saying, "3, so it's 30." Lilly starts to check the measurement and counts by ones. Melissa picks this count up, but counts (as did Emily and Janet) each group of ten by ones (1, 2, 3 . . . 10, 1, 2, 3 . . . 10, etc.) Melissa rechecks her answer and counts by tens, touching the tenth cube in each group as she counts, "10, 20, 30." She checks it one more time and counts the stack by ones, but even in this count, her emphasis, both physical and verbal, is on the decade in each grouping.

In the second part of Clip 7, the girls are working with the long side of the paper. They use the same color and grouping of cubes and count in a similar manner (first by tens, 10, 20, 30, 40 and counting on the units, 41, 42, 43, 44; and then recheck the entire stack by counting by ones).

LINDSEY AND LUCERO—CLIP 5

These girls also use two colors (green and red) of cubes grouped in ten. Lindsey silently counts the groups of ten, touching each group. Lucero also skip counts the cubes by ten ("10, 20, 30") and counts on the units, "31, 32, 33, 34." She misses a stack of ten, which is hidden under the clipboard. Lindsey corrects her count, by including the hidden stack of cubes, "10, 20, 30, 40, 41, 42, 43, 44, 45," and gets an additional unit. It is interesting to note on their recording sheet, they've marked "44" as the measurement.

Appendix B

HANDY GUIDE TO THE CD-ROM CLIPS

JOURNEY 1

Folder	Page	Clips
The Class at Work	Introduction	
	Developing the Context	1
	Anticipating Strategies	
	Children at Work	
	Alexander & Haley	6
	Amirah & Jonathan	9
	Andrew & Shannon	4
	Angelica & Diana	72, 74, 75, 73
	Anthony & Sarah	3
	Emily & Janet	10
	Fernando & Josué	2
	Francisco & Julio	
	Lilly & Melissa	7
	Lindsey & Lucero	5
	Looking Back	
	Reflections on Children at Work	
	What Would You Do Next?	
	Building the Blueprint	
	The short side of the blue paper	13
	The long side of the blue paper	15
	The long side of the purple paper	18
	The long side of the cardboard	19
	The very long side of the chart paper	21
	The long side of the yellow paper	22
	Where is 66?	23, 24, 26, 27, 28
	What to mark?	29
	Where to place 80?	30
	Where to place 60?	31
	The Next Day: Mental Math	51, 53, 54, 55, 56
	Several Days Later: Mental Math	66
	Backburner: A Moment for Further Reflections	

37

Appendix B

Folder	Page	Clips
Growth and Development	Investigating Learning	
	Alexander	6
	Amirah	9, 16, 55
	Andrew	4
	Angelica	72, 75, 74, 73, 32, 54
	Anthony	3
	Diana	72, 75, 73, 74
	Emily	10, 33, 28, 43, 36, 48, 56
	Fernando	2
	Francisco	
	Haley	6, 40, 44
	Janet	10, 45, 31, 71
	Jonathan	9, 35
	Josué	2, 34, 37, 47
	Julio	50
	Lilly	7
	Lindsey	5
	Lucero	5, 26, 41, 46
	Melissa	7, 20, 39, 70
	Sarah	3, 22
	Shannon	4, 27, 42
	Backburner: A Moment for Further Reflections	

JOURNEY 2

Folder	Page	Clips
The Role of Context	Introduction	
	Hildy's Use of Context	
	Selecting Moments as Evidence	6, 10, 11, 52, 66
	Building a Learning Environment	
	Choosing the Numbers	11
	Using Context to Scaffold Learning	
	The Role of Context	6, 11, 52, 66
	Backburner: A Moment for Further Reflections	
The Role of the Teacher	Introduction	
	Supporting Development	6, 11, 52, 66
	Maximizing a Moment	29
	Posing Important Questions	6, 11, 52, 66
	Facilitating Dialogue	17
	Pair Talk	29, 6, 11, 52, 66
	Providing Think Time	67, 52
	Backburner: A Moment for Further Reflections	
Developing a Community	Introduction	
	Evidence of a Community	6, 11, 52, 66
	Building and Supporting a Community	6, 11, 52, 66
	Backburner: A Moment for Further Reflections	
The Number Line	Interview with the Teacher	68
	Representing Emergent Modeling	6, 11, 52, 66, 68
	Backburner: A Moment for Further Reflections	

Appendix C

DIALOGUE BOXES

DIALOGUE BOX A: *KID WATCHING*

The following discussion occurred in the second session working with the CD-ROM, *Working with the Number Line, Grade 2: Mathematical Models.* Before sending participants off to examine *Children at Work*, the facilitator used an LCD to watch Alexander and Haley with the whole group to develop their kid watching.

> *Facilitator:* We've just watched Alexander and Haley at work. Think about what the students are doing and saying and see if you can describe their strategies. I'll ask some people to share. Before we start, I'd like to set some parameters for our discussion. Let's try to just get our ideas on the table—I'll record them up here on the LCD so everyone can see what's been said. For now, I'd like us *not* to comment on what other people are saying—even if you strongly disagree. Feel free to repeat what's been said if you think it's really important. At the end of the discussion, we'll look at our list and see what ideas people had in common and what ideas are different.
>
> *F:* (*to a participant*) What is your story?
>
> *Participant:* Haley is counting the cubes by ones.
>
> *F:* (*to another participant*) What's your story—and let's keep going around this table, all of you share your ideas.

Participants' comments include ones specific to what the students are doing (*Haley is counting by ones; he's counting by fives*), judgments about students' abilities (*Alexander is the stronger student; he's helping Haley count*), interpretations of their emotions (*Haley is getting very frustrated*), and questions about Hildy's planning (*I kept wondering why Hildy paired them. Is it because one student is stronger than the other? Alexander certainly seems to be providing a lot of help to Haley*).

> *F:* So far we have very different stories. How can this be? Can stories be so different and each be true? (*reads from the chart*) "Haley is counting by ones; Haley is counting by fives; Haley is counting by tens." How do you count by ones *and* count by fives? Is this possible? Let's watch the clip again and see if we can start to agree on what's happening. Let's also try to only say things we can find evidence for in the clip, so talking about Hildy or what the students are feeling—there's no way

to support this—these are interpretations. (*Second viewing of the video clip*)

P: Haley counts by ones and touches each cube as he counts.

P: Both boys are counting; I couldn't be sure who was counting by ten. The first time I thought it was Alexander, now I'm not so sure. It might be both of them.

P: Haley counts by ones, but he keeps track of his tens. You can see it at the end . . . in the last clip, each red cube marks a ten.

P: He only got that strategy because Alexander gave it to him. Otherwise he would have kept counting by ones.

F: For now I'd like to keep us focused on *what's actually happening*—the physical actions and words in the clips. We don't know if Alexander gave him the strategy; that's an interpretation. You keep talking about how Haley counts. How does he count?

P: By ones.

P: Is Haley counting by ones? People keep saying that, but I'm not so sure he is.

F: I'd like people to think about what Jamie just said for a moment. Is Haley counting by ones? Talk to the person next to you about this: How is Haley counting?

The facilitator listens to different conversations and then asks one pair to share.

P: Well, we were disagreeing at first. I said he counts by ones, but Mary said he didn't. But she convinced me when she pointed out that he never counts beyond five. He says, "1, 2, 3, 4, 5," but there's no 6. So he's counting by ones, but he's not counting by ones.

F: I'm not sure I understand what you mean. Did someone else understand what Dee is talking about? If you did, could you paraphrase what she just said and help me understand?

P: It sounded like she said the answer is yes and no. Haley is counting by ones and he isn't counting by ones. He counts by ones to five, so in that way he's counting by ones, but, but since he never says 6, he isn't really counting by ones' cause then he'd keep going, 7, 8, 9 . . .

P: That's my point exactly! You nailed it.

P: We agree with Dee's group—we talked about the same thing. Haley keeps counting 1, 2, 3, 4, 5, 1, 2, 3, 4, 5. We talked about how he's only using white cubes so if his strategy was to count the groups, there are no groups to count. It's almost like he's forced to do that—count to five to help him keep track.

P: That makes sense, we talked about that too, how he counted. 1, 2, 3, 4, 5, 1, 2, 3, 4, 5, but we couldn't figure out why he was doing that.

F: So a number of you have commented on how Haley is counting to five and says, 1, 2, 3, 4, 5, 1, 2, 3, 4, 5. Did anyone hear anything different? Hmm. That's interesting—everyone heard Haley count the same way. Can anyone in here show us his actions as he counts?

There is a heated discussion on *how* Haley counts; one participant says, "I think I saw him use both hands," but is not able to imitate his actions for the group. Not everyone agrees that Haley is using two hands as he counts.

F: I'm going to show the clip again. I want you to focus on: What is Haley doing? Can you imitate his strategy? Then I'd like you to think about this: if Haley had a voice in his head, not speaking aloud, what would that voice be saying as he counts? What is he thinking? What thinking underlies Haley's actions?

After the third viewing of the video clip, the facilitator asks the participants to talk in pairs to see if they can agree on how Haley counted and what is going on mathematically in Haley's mind to be able to use such a strategy. After pairs talk, the participants are brought back together for a whole-group discussion.

> P: We talked about how easy it is to judge a student without really taking into account what thought is behind his actions. I really thought Haley was counting by ones. That's where I stopped—at that judgment. Now I think there's a lot more going on there than I had first realized. It's such a complicated strategy.
>
> F: What makes his strategy so complicated?
>
> P: Well, that he has so many things happening at the same time. He's keeping track of a lot of information. He's counting by ones with one finger, but marking the fives with the other. When he gets two groups of fives, he says "10." He keeps track of how many tens, by skip counting—you hear him say, "10, 20, 30."
>
> F: Such a complicated strategy. He has a finger that counts, and a finger that moves—Haley has to do so many things at the same time.
>
> P: I think that's why his strategy changed the way it did. In the end, the pink cubes that mark the groups of ten, those were the tens he was counting out before when he said, 1, 2, 3, 4, 5, 1, 2, 3, 4, 5, 10, 1, 2, 3, 4, 5, 1, 2, 3, 4, 5, 20.
>
> P: Amazing! I didn't see it earlier, but now that you're describing it—he has two systems going at once. One hand is keeping track of the fives and tens, the landmarks. The other hand is counting by ones to five. And then later on he marks the groups of ten with another color, and counts just the groups.
>
> F: Same video—so many different stories!! So many questions about Haley, what he knows, what his strategy means. As you go off to work now with the clips of other children in Hildy's classroom, I'd like you to remember the way we worked together this morning. See if you can tell the story—stay away from judging students, and making interpretations. See if you can imitate their actions. Think about, how is what they're saying connected to what they're doing? Once you can accurately describe the students' strategies start to think about what does the student have to know mathematically to be able to use that strategy.

FACILITATOR NOTES ON DIALOGUE BOX A

This is an example of an interactive learning environment in which the facilitator works on a number of levels. He builds a community of learners by

- valuing each participant's thoughts—note the initial recording of each participant's ideas without commenting on what they are saying;
- using pedagogical tools like pair talk *to foster and encourage communication between participants*;
- encouraging participants to comment on each other's ideas (the ball of conversation is passed not just between teacher and participant and back again to the teacher, but between participants).

The facilitator also works to sharpen participants' powers of observation, one of the primary goals of a beginning journey. The facilitator

- **controls the flow of the conversation** (it doesn't jump all over the place). Initially this is done by asking the participants, *What is your story?* and not allowing participants to comment on each other's ideas.

- **sets the parameters for kid watching.** This is done slowly and subtly; a list of rules is *not given* to participants at the outset, *but evolves from the contradictions that arise from the different stories they tell and the need for accurate retelling;*
- **develops participants' kid watching** by asking them for supporting evidence for their interpretations and observations;
- **creates disequilibrium** when their stories don't mesh and asks, *How can this be? Can stories be so different and each be true?*

The facilitator uses the technology as a tool to

- **deepen thinking.** The repeated viewing of the clips helps participants refine their thinking and reflect on the ideas of others. The level of accuracy in their stories improves as the facilitator pushes participants toward a more precise retelling (e.g., *Can you show us how he counted?*) With each viewing the facilitator's questions become more focused on detail. At the end of the discussion, participants are sent back to work in pairs, expected to use some of the kid watching tools they have just developed. Witness the use of cycles of observing, discussing, analyzing, reflecting, developing narrative knowledge, and expanding personal repertoires and generalizing (as described in the *Overview Manual*).
- **work with the range of learners in the class, and provide experiences broad enough to meet different needs.** (e.g., Some participants will need to spend more time developing nonjudgmental ways of looking at students than others. For participants who quickly develop their powers of observation, *another kind of journey*, one in which they begin to think about the mathematical ideas that underlie students' actions [strategies] and words is more appropriate.)

DIALOGUE BOX B: *BUILDING THE BLUEPRINT*

The following conversation was taken from a three-day workshop using the CD-ROM, *Working with the Number Line, Grade 2: Mathematical Models.* After participants had worked through *Building the Blueprint* on the CD-ROM and answered the corresponding questions, the facilitator brought them together as a whole group to begin to think about Hildy's number choice, which, in later discussion would be connected to her teaching plan.

Facilitator: Hildy is working with the whole group. Let's talk about Hildy's number choice. What is the effect of her number choice on their thinking? What were the numbers she used? Make a list of the numbers you remember.

Participants make lists, but quickly realize that their lists do not match. The facilitator suggests they recheck and together they construct the following list:

10
14
22
35
46
84
66
80
60

F: Was the number choice a coincidence? Any reason she used these numbers?

Participants discuss the connection Hildy's number choice has to five and ten as landmarks. Comments are also made about *how* Hildy got these numbers to come up as measurements.

Participant: I don't think Hildy measured out paper.

F: Why might she do it (*measure out the paper*)? Are these measurements a coincidence?

P: First paper, starting at 10 was purposeful.

F: Why?

P: I think the purpose was to get them to unitize fives or tens, push kids to unitizing.

In the ensuing conversation, participants think about why Hildy *chose* certain numbers to work with and how she is structuring her discussion. Comments include *working with unitizing, working with five and ten as a landmark*. One participant says, "I think 66—she really thought about the conversations they might have around that."

P: I think she made a choice . . . that she had paper with 10, 20, 30, 40 to scaffold moving to 84.

F: She *did* cut it.

Many participants are amazed that Hildy precut the papers to ensure certain numbers would come up.

F: Let's continue looking at her decision making. Was that train of cubes in the classroom or not? How do the children use it?

Then participants share their findings on how each number was placed on the paper strip (e.g., for 22, a teacher said, "She said, (*Amirah*) 'Two tens and then count by ones to 22.' ").

F: All these strategies. Are they the same? Different? Related? What's the connection to what the kids did in measuring and what they're doing as they build the blueprint?

P: Using the structure of fives.

P: Counting by fives and adding on.

P: Going back to kids at work—they're seeing it more now; they're using fives and tens, when before they were counting by ones.

F: How did she do that? What does she do to make that happen—or is it a coincidence?

P: Hildy puts the cubes way up on the board so they can't touch them.

P: When I think back over the whole investigation, the difference—first they work with the cubes to measure, but now in this discussion, she doesn't let them touch them—the cubes are high, the kids can't tag—they have to use the visual.

P: Cubes are a way to scaffold their thinking—color helps them visualize. Hildy moves it from concrete to abstract.

P: The structure is imposed. They have to think about two groups of five.

F: Several people talked about concrete to abstract. For yourselves, for a moment, think about that order from concrete to abstract. Does this investigation move from very concrete to very abstract?

P: Using manipulatives, the cube line, using a pointer, showing students this is this many, she could pull cubes away, connect numbers she's writing to the cube line—that's concrete.

P: Two color cubes—concrete, paper of different sizes—concrete, abstraction was how they grouped, how they counted.

P: Physical actions and mental actions are simultaneous. I think about where is 22 mentally, but it's represented physically with the white and green cube line. Hildy used cubes to prove the relationships, where is 66—that's concrete.

P: She started with them putting cubes together in trains. They used her cube train to mark the open number line she was building. Students used her number line, filled it in with measurements. The measurements were physical; the relationships between them weren't.

F: Seems like we can merge all three. They work with cubes, look at cubes, but can't touch them—that's a constraint. Students used two cubes, had a kind of cube line, but Hildy used her cube line differently. They had a line with no numbers; they built it. What's the effect of that? How does it influence student thinking?

P: Made the students think about the number line in their head.

P: Emily was ready to use the number line as a tool.

F: Someone else said Emily was using the number line as a tool. Think about this. Is there proof that Emily was using this as a tool?

The discussion here focused on Emily's shift in thinking. Participants noted

- how Emily was using the cube line in *Building the Blueprint*, (always starting at the beginning and counting up to the number she was placing);
- the point at which her strategy shifted (with Hildy's question, *Can you do it without going back to the beginning?*); she takes *two jumps of ten* to get from 46 to 66.

F: What things are supporting development? What did Hildy do to help kids develop this strategy? Make a list with your partner; we'll come back together in a few minutes.

Participants created lists that were shared in a whole-group discussion. During the discussion the facilitator wrote their comments on chart paper. The facilitator organized what they were saying into three separate groups: Hildy's Use of Number, Manipulatives, and Constraints.

Hildy's Use of Number	Manipulatives	Constraints
• Numbers chosen deliberately • Order of number • Recording the numbers to used to build the blueprint emphasize landmarks • Changes the numbers to make it harder (stretch kids' thinking)	• Multi-link cubes, two colors only • Cube line with two colors • Organized in groups of five • Cube line can be manipulated to show thinking • Paper strip below cube to record measurements on	• Kids can't touch the cubes • Precut papers students measured are not physically used to build the blueprint • Hildy records the numbers as children describe where they go

The facilitator reemphasized the comment that there was a shift in student strategies when they were working with where is 66 in relationship to 46 and 84:

F: Starting at a convenient number, the number line works as a tool—they can make nice jumps of 10. It's one of the moments where students start to use the number line as a tool. Hildy did a lot to make learning happen. Hildy built in constraints—made it just a little bit harder. In a way, she scaffolded growth and development. She enabled the students to use that group of five. Tomorrow, we'll go further. Tonight, in your journals, I would like you to think about how Hildy's choices influenced what the kids did. How did their strategies change? If a strategy changed, what made that happen? What direct and indirect things did the teacher do to help the kids grow mathematically?

DIALOGUE BOX C: STUDENTS' GROWTH AND DEVELOPMENT OVER TIME

One facilitator worked with the written assignment about the growth and development of a child in this way: In class, she invited participants to choose one of the following children—Amirah, Angelica, Emily, or Janet—to write their impressions (what they remembered) about the child from *Children at Work*. Here are some questions that she gave to help them with their remembering:

- What mathematical ideas has this child constructed?
- What evidence do you have to support this?
- How did the child's strategies change?
- What facilitated growth and development?
- If you do not have a clear picture of this child or cannot remember details, do you now have any questions?

When the participants finished writing their impressions, the facilitator had them group themselves according to the child they picked. She gave them some time to read their impressions to each other and to take notes on conflicting impressions, consistent impressions, and the questions that they still had. Then she called all the groups together and charted their notes as each of the four groups shared.

Facilitator: Let's just look at Janet briefly. Many of you agreed that (*reads from their chart*) "She added 43 and 20 by splitting the 20 into two tens and taking jumps of tens to get to 63," but disagreed on other details (*continues reading*):

"Emily did all the work in their pair; Janet just followed along silently."

"Janet was a silent, but equal partner. She used the strategy Emily started—counting out ten—that means she understood it."

"She develops ten as a landmark. She used it in her measurement with Emily and she used it when she solved 43 + 20 by adding two tens to 43."

"She's all over the place; she's stronger with addition than with subtraction."

"She struggled with subtraction."

"She needed help figuring out where to mark 80—how many cubes to go back from 84."

"She worked with place value when she knew that 66 was either 6 back or 6 forward from 60."

"She had difficulty knowing where to mark 80 in relationship to 84, but used what she had learned to solve where to mark 60 in relationship to 66."

So there were a number of differences in how you remembered Janet. Some of your impressions of her are more detailed than others. But there was strong disagreement as to her role as Emily's partner in "The Class at Work" that ranged from her having no strategy to her strategy as an example of her developing ten as a landmark.

This is your assignment. With your written impression as an entry point, go into the "Growth and Development" folder in *Journey 1*, the section where the names of the children are listed. You can use the video clips here as a database to help you create an in-depth portrait of your child's growth and development. Remember to find evidence to support what you are saying. A statement like, *Janet does not know how to subtract* needs to have supporting details. What are her struggles? Do her struggles change? If so, how? You can use the questions I gave you earlier to help you with this.

DIALOGUE BOX D: *HILDY'S TEACHING PLAN*

The facilitator began the session by asking each pair/triad to finish working with *Building the Blueprint* and also to look at Hildy's two minilessons. After giving them forty minutes to work on this, participants were asked to think about Hildy's plan and to *compile a list of key elements in her plan* that they would like to share with the group. As participants discussed this, the facilitator listened in to each group to get a general sense of how the students were interpreting Hildy's overall plan and also to see if there were any misconceptions. Each group was given chart paper and asked to record their compiled list. Of these listings, two groups were picked to share in the whole-group discussion.

The first group shared the chart shown below:

What Hildy Did	How It Helped
Put out two colors of multilinks for students to use as they measured	Students created groups with the colors to count them
Used two colors of cubes in the cube line with the constraint that the students couldn't touch them	Forced students to use the groups to count; moved them away from counting by ones to skip counting by fives or tens
Used two groups of five in the cube line	Helped develop the idea of unitizing
Used cubes and measurements to build number line	They built the number line together and the discussion helped kids place numbers in relationship to each other
Recorded landmarks differently from numbers on the paper strip	Helped students think about ten as a landmark number (14 can be thought of as 10 + 4)
Removed all the concrete materials Created minilessons around using ten	Began to use the number line as a tool Students worked with tens or *friends of ten* to solve the problems (43 + 20 = 43 + 10 + 10)

The second group shared their plan:

Dialogue Box D:
Hildy's Teaching Plan

MATERIALS:	HILDY'S PLAN
• cubes (two colors)	students can use colors to make groups to help them count
• precut paper	helps her get the measurements she wants on her number line; she uses these numbers to build her number line (10, 22, 35, etc.)
• recording sheet for students	when students come to whole-group share, they have their measurements on their clipboard (saves time, good management tool because the sheet is organized so that when she says, the short side of the purple paper, students can easily locate it on their recording sheet)
• cube line with two colors of cubes grouped in fives	helps students develop different strategies (skip counting, counting on), helps them develop unitizing; Hildy knows this organization (the grouping) will help students place the measurements on the paper strip; they can use either five or ten
• paper strip underneath the cube line	Hildy records the measurements, but *plans* how she'll record them (she records landmarks larger than other numbers)

Other important things Hildy did:
(What they wrote on chart paper)

• started with a rich context problem

What the participants said:
Hildy knew she could come back to the context when students got confused (where should 66 go on or off the cube—when she went back to the measurement context, they suddenly knew 66 was not on the cube).

• number choice

Hildy carefully selected the numbers and changed how she used them. 46 was the point where she changed the flow—it was purposeful because she knew 46 would help kids figure out where's 66, but now they wouldn't necessarily be working with landmarks like ten, twenty . . . but thinking with groups of ten.

• used the idea of working with tens (making tens, going to a ten) in her minilessons

• moved students from concrete to abstract

Hildy worked with a physical and visual representation of number (when they built the number line) to a mental one (the last string where she did not use the cube line or paper strip). She moved kids away from using the cube line so that they would start to work with numbers in their heads.

DIALOGUE BOX E: *PAIR TALK*

In an initial conversation, before sending participants off to work with this section, a facilitator asked them how they used pair talk in their own teaching. She made of list of participants' comments to return to after they had worked with *Facilitating Dialogue*. Participants then explored the selected and annotated clips of Hildy's use of pair talk and thought about places where they might have used it in the whole-group discussions. The dialogue below is taken from their conversation on pair talk.

Facilitator: Many of you in your small groups commented on how infrequently Hildy used pair talk and how surprised you were by this.

Participant: We spent a lot of time talking about this in our group, about why we use pair talk and about the way Hildy used it. We think she could have used it more frequently. There were a lot of kids that needed to talk more, that seemed to be very quiet.

F: So one way to use pair talk is to help students who are relatively quiet during the whole-group discussion?

P: Absolutely. I think you can use pair talk to build community too—that's how I use it in my classroom. There were lots of instances where Hildy could have had pair talk to do that.

P: I would have used pair talk more when kids were sharing their strategies for placing numbers on the blueprint.

F: Can you give us an example of what you mean—or where you would have used pair talk?

P: Well, I thought Josué's strategy was pretty complicated and I bet there were a number of kids who had no idea what he was doing. I would have asked the other students to talk to their partners about his strategy and then had one of the partners report back to the group what they had said.

F: Can you help us understand why you picked Josué's strategy and not, say, Amirah's?

P: His strategy seemed like a pretty important one—he's talking about making *jumps of ten*—putting two fives together to make ten, and working with ten to locate a number. I think it would have been a good opportunity to get kids talking about his strategy because I'm not sure how many of them understood what he was doing. And since using ten is a landmark strategy, it's important for them to know.

F: So if we think about some of the ways we've been talking about pair talk, it's been to bring more reticent learners into the conversation; to build community; and to understand strategies that might be mathematically challenging. I'd like to talk about Hildy's use of pair talk. Think about *how and when* Hildy used pair talk. Are there any other ways Hildy used pair talk than the ways we've mentioned using it?

P: Our group was really struck by how infrequently Hildy used pair talk—you know, the idea that less is more. And we came up with one other way she used it that's a very different way than what's been said so far. She used it when kids were struggling with a huge mathematical idea.

F: Can you give us an example of what you mean?

P: There were two we came up with: Where is 66?, when there was the potential for kids to use the numbers up on the blueprint to place a number . . .

F: And why was that a big moment mathematically?

P: Because the blueprint was becoming a number line.

P: They were working with number relationships to place it. I mean when you think of the big things that happened at that moment: Haley models the relationship with his hands . . .

F: Models number space?

P: Yeah, when he says somewhere here is 50 and here is 70 and 60 is right in the middle. That was a really big moment.

P: And it also helped Emily move away from going back to 0 to place a number. That was a big moment in the class too.

P: So what was your other example—you said there were two?

P: Where to mark 66, at the edge or on the cube; we thought that was a big mathematical moment too. Kids are now thinking about a number as distance, the whole context of measurement built up to that point. I thought it was incredibly powerful.